SERIES EDITOR: ALAN SMITH

Modular Maths for Edexcel

Mechanics 2

Second Edition

DAVID O'MEARA
PAT BRYDEN, JOHN BERRY, TED GRAHAM

Hodder & Stoughton
A MEMBER OF THE HODDER HEADLINE GROUP

M2

Acknowledgements

We are grateful to the following companies, institutions and individuals who have given permission to reproduce copyright material in this book. Every effort has been made to trace and acknowledge ownership of copyright. The publishers will be glad to make suitable arrangements with any copyright holders whom it has not been possible to contact.

Photos:
p1 © John Greim/Science Photo Library
p19 © Action-Plus Photographic
p51 © Glyn Kirk/Action-Plus
p74 Mc Esher *Waterfall* © 2000 Arts B.V. – Baarn – Holland, all rights reserved
p95 © Neil Tingle/Action-Plus

Orders: please contact Bookpoint Ltd, 130 Milton Park, Abingdon, Oxon OX14 4SB.
Telephone: (44) 01235 827720, Fax: (44) 01235 400454.
Lines are open from 9.00–6.00, Monday to Saturday, with a 24 hour message answering service.
You can also order through our website www.madaboutbooks.co.uk.

British Library Cataloguing in Publication Data
A catalogue record for this title is available from The British Library

ISBN 0 340 885289

First published 2000
Second edition published 2004
Impression number 10 9 8 7 6 5 4 3 2 1
Year 2010 2009 2008 2007 2006 2005 2004

This work includes material adapted from the MEI Structured Mathematics series.

Cover photo from The Image Bank/Getty Images
Typeset by Tech-Set Ltd, Gateshead, Tyne & Wear.
Printed in Great Britain for Hodder & Stoughton Educational, a division of Hodder Headline Plc, 338 Euston Road, London NW1 3BH by Martins the Printers Ltd.

Edexcel Advanced Mathematics

The Edexcel course is based on units in the four strands of Pure Mathematics, Mechanics, Statistics and Decision Mathematics. The first unit in each of these strands is designated AS, and so is Pure Mathematics: Core 2; all others are A2.

The units may be aggregated as follows:

3 units	AS Mathematics
6 units	A Level Mathematics
9 units	A Level Mathematics + AS Further Mathematics
12 units	A Level Mathematics + A Level Further Mathematics

Core 1 and 2 are compulsory for AS Mathematics, and Core 3 and 4 must also be included in a full A Level award.

Examinations are offered by Edexcel twice a year, in January (most units) and in June (all units). All units are assessed by examination only; there is no longer any coursework in the scheme.

Candidates are not permitted to use electronic calculators in the Core 1 examination. In all other examinations candidates may use any legal calculator of their choice, including graphical calculators.

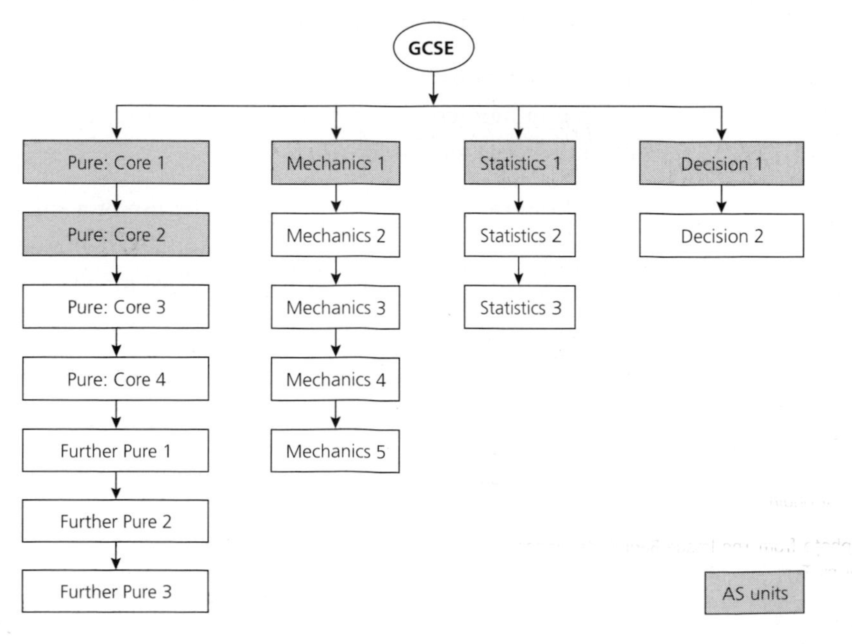

INTRODUCTION

This is the second book in a series written to support the Mechanics units in the Edexcel Advanced Mathematics scheme.

Throughout the series the emphasis is on understanding and applying a wide variety of mathematical skills, and particularly on the application of mathematics to modelling.

There are six chapters in this book. They are presented in the same order as the specification, although many teachers will wish to study them in a different order. Cross-references have been reduced to the minimum to make this possible. Full answers are included.

The first chapter deals with projectiles when air resistance is ignored. It finishes by finding and using the general equation of the trajectory. Kinematics is also the subject of chapter two, which introduces the use of calculus.

Chapter three deals with finding centres of mass by non-calculus methods for two-dimensional objects only. Finding the centre of mass of frameworks is included.

In chapter four you meet and use energy, work and power for both one- and two-dimensional motion. Chapter 5 builds on the work in *Mechanics 1* on momentum, this time extending the work to two dimensions and adding the use of Newton's law of restitution for direct impact only.

The last chapter again builds on work from *Mechanics 1*, this time extending the principle of moments to include forces at angles other than 90°. This is where you will find the classic ladder problem.

Throughout the book you will require the use of a calculator. Computer packages and graphics calculators may be helpful in clarifying new ideas, but you should remember that certain calculator restrictions may be enforced by the examination board.

I would like to thank the many people who have helped in the preparation and checking of material. Special thanks to Pat Bryden, John Berry and Ted Graham.

David O'Meara

CONTENTS

Chapter one

PROJECTILES

Swift of foot was Hiawatha;
He could shoot an arrow from him,
And run forward with such fleetness,
That the arrow fell behind him!
Strong of arm was Hiawatha;
He could shoot ten arrows upwards,
Shoot them with such strength and swiftness,
That the last had left the bowstring,
Ere the first to earth had fallen!

The Song of Hiawatha, *Longfellow*

Look at the water jet in the picture. Every drop of water in a water jet follows its own path which is called its *trajectory*. You can see the same sort of trajectory if you throw a small object across a room. Its path is a parabola. Objects moving through the air like this are called projectiles.

MODELLING ASSUMPTIONS FOR PROJECTILE MOTION

The path of a cricket ball looks parabolic, but what about a boomerang? There are modelling assumptions which must be satisfied for the motion to be parabolic. These are

- a projectile is a particle
- it is not powered
- the air has no effect on its motion.

EQUATIONS FOR PROJECTILE MOTION

A projectile moves in two dimensions under the action of only one force, the force of gravity, which is constant and acts vertically downwards. This means that the acceleration of the projectile is $g\,\text{ms}^{-2}$ vertically downwards and there is no horizontal acceleration. You can treat the horizontal and vertical motion separately using the equations for constant acceleration.

To illustrate the ideas involved, think of a ball being projected with a speed of $20\,\text{ms}^{-1}$ at $60°$ to the ground as illustrated in figure 1.1. This could be a first model for a football, a chip shot from the rough at golf or a lofted shot at cricket.

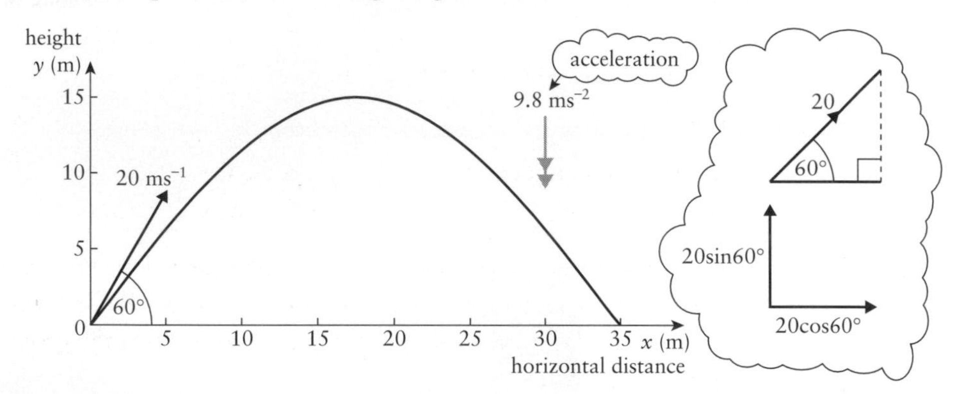

FIGURE 1.1

Using axes as shown, the components are

	Horizontal	Vertical
Initial position	0	0
Acceleration	$a_x = 0$	$a_y = -9.8$

This is negative because the positive y axis is upwards

	Horizontal	Vertical
Initial velocity	$u_x = 20\cos 60°$	$u_y = 20\sin 60°$
	$= 10$	$= 17.32$

Using $v = u + at$ in the two directions gives the components of velocity.

Velocity

	Horizontal	Vertical
	$v_x = 20\cos 60°$	$v_y = 20\sin 60° - 9.8t$
	$v_x = 10$ ①	$v_y = 17.32 - 9.8t$ ②

$a_x = 0 \Rightarrow v_x$ is constant

Using $s = ut + \frac{1}{2}at^2$ in both directions gives the components of position.

Position

	Horizontal	Vertical
	$x = (20\cos 60°)t$	$y = (20\sin 60°)t - 4.9t^2$
	$x = 10t$ ③	$y = 17.32t - 4.9t^2$ ④

You can summarise these results in a table.

	Horizontal motion		Vertical motion	
Initial position	0		0	
Acceleration	0		−9.8	
Initial velocity	$u_x = 20\cos 60° = 10$		$u_y = 20\sin 60° = 17.32$	
Velocity at time t	$v_x = 10$	①	$v_y = 17.32 - 9.8t$	②
Position at time t	$x = 10t$	③	$y = 17.32t - 4.9t^2$	④

The four equations ①, ②, ③ and ④ for velocity and position can be used to find several things about the motion of the ball.

What is true at

(a) the top-most point of the path of the ball?

(b) the point where it is just about to hit the ground?

When you have decided the answer to these questions you have sufficient information to find the greatest height reached by the ball, the time of flight and the total distance travelled horizontally before it hits the ground. This is called the *range* of the ball.

The maximum height

When the ball is at its maximum height, H m, the *vertical* component of its velocity is zero. It still has a horizontal component of $10\,\text{m s}^{-1}$ which is constant.

Equation ② gives the vertical component as

$$v_y = 17.32 - 9.8t$$

At the top $0 = 17.32 - 9.8t$

$$t = \frac{17.32}{9.8} = 1.767$$

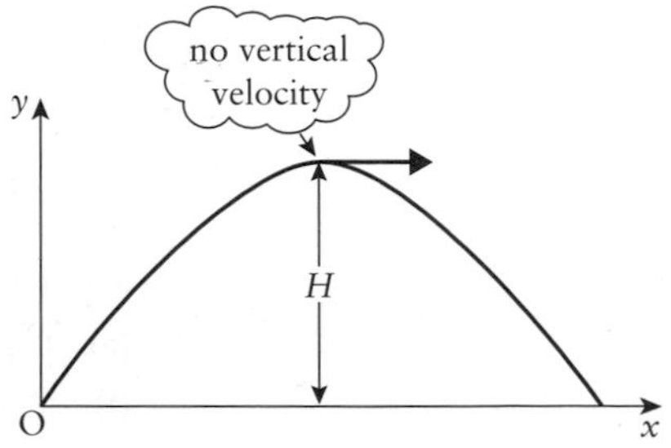

FIGURE 1.2

To find the maximum height, you now need to find y at this time. Substituting for t in equation ④,

$$y = 17.32t - 4.9t^2$$
$$y = 17.32 \times 1.767 - 4.9 \times 1.767^2$$
$$= 15.3$$

The maximum height is 15.3 m.

The time of flight

The flight ends when the ball returns to the ground, that is when $y = 0$. Substituting $y = 0$ in equation ④,

$$y = 17.32t - 4.9t^2$$
$$17.32t - 4.9t^2 = 0$$
$$t(4.9t - 17.3) = 0$$
$$t = 0 \text{ or } t = 3.53$$

Clearly $t = 0$ is the time when the ball is thrown, so $t = 3.53$ is the time when it lands and the flight time is 3.53 s.

The range

The range, R m, of the ball is the horizontal distance it travels before landing.

R is the value of x when $y = 0$.

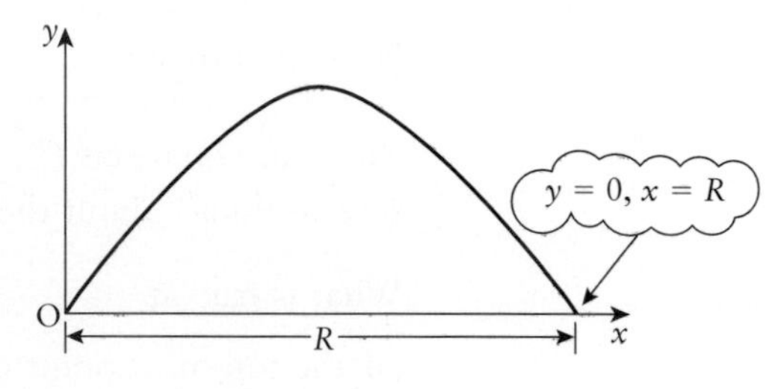

FIGURE 1.3

R can be found by substituting $t = 3.53$ in equation ③: $x = 10t$. The range is $10 \times 3.53 = 35.3$ m.

EXERCISE 1A

In this exercise take upwards as positive and use 9.8 ms^{-2} for g unless otherwise stated. All the projectiles start at the origin.

1 In each case you are given the initial velocity of a projectile.

(i) Draw a diagram showing the initial velocity and path.
(ii) Write down the horizontal and vertical components of the initial velocity.
(iii) Write down equations for the velocity after time t seconds.
(iv) Write down equations for the position after time t seconds.

(a) 10 ms^{-1} at 35° above the horizontal.
(b) 2 ms^{-1} horizontally, 5 ms^{-1} vertically.
(c) 4 ms^{-1} horizontally.
(d) 10 ms^{-1} at 13° below the horizontal.
(e) U ms^{-1} at angle α above the horizontal.
(f) u_0 ms^{-1} horizontally, v_0 ms^{-1} vertically.

2 In each case find

(i) the time taken for the projectile to reach its highest point

(ii) the maximum height.

(a) Initial velocity $5\,\text{ms}^{-1}$ horizontally and $14.7\,\text{ms}^{-1}$ vertically.

(b) Initial velocity $10\,\text{ms}^{-1}$ at $30°$ above the horizontal. Use $10\,\text{ms}^{-2}$ for g.

3 In each case find

(i) the time of flight of the projectile

(ii) the horizontal range.

(a) Initial velocity $20\,\text{ms}^{-1}$ horizontally and $19.6\,\text{ms}^{-1}$ vertically.

(b) Initial velocity $5\,\text{ms}^{-1}$ at $60°$ above the horizontal.

PROJECTILE PROBLEMS

When doing projectile problems, you should treat each direction separately. Start with the direction that you know most about.

EXAMPLE 1.1

A ball is thrown horizontally at $5\,\text{ms}^{-1}$ out of a window 4 m above the ground.

(a) How long does it take to reach the ground?

(b) How far from the building does it land?

(c) What is its speed just before it lands and at what angle to the ground is it moving?

Solution The diagram shows the path of the ball. It is important to decide at the outset where the origin and axes are. You may choose any axes that are suitable, but you must specify them carefully to avoid making mistakes. Here the origin is taken to be at ground level below the point of projection of the ball and upwards is positive. With these axes, the acceleration is $-g\,\text{ms}^{-2}$.

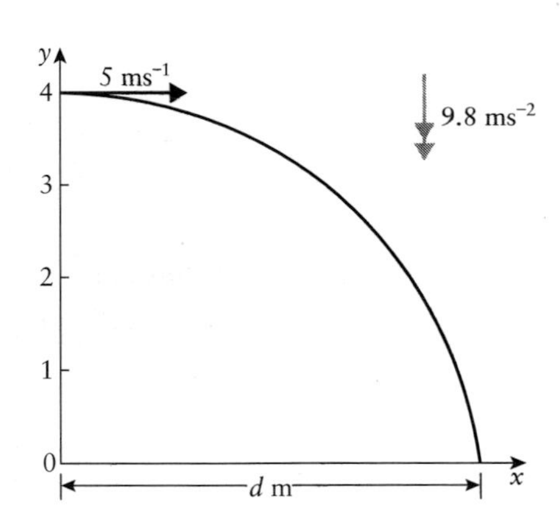

FIGURE 1.4

Position: Using axes as shown and $s = s_o + ut + \frac{1}{2}at^2$ in the two directions

Horizontally $\quad x_o = 0,\ u_x = 5,\ a_x = 0$

$$x = 5t \qquad ①$$

Vertically $\quad y_o = 4,\ u_y = 0,\ a_y = -9.8$

$$y = 4 - 4.9t^2 \qquad ②$$

(a) The ball reaches the ground when $y = 0$. Substituting in equation ② gives

$$0 = 4 - 4.9t^2$$
$$t^2 = \frac{4}{4.9}$$
$$t = 0.904$$

The ball hits the ground after 0.9 s.

(b) When the ball lands $x = d$ so, from equation ①, $d = 5t = 5 \times 0.904 = 4.52$. The ball lands 4.52 m from the building.

Velocity: Using $v = u + at$ in the two directions

Horizontally $\quad v_x = 5 + 0$

Vertically $\quad v_y = 0 - 9.8t$

(c) To find the speed and direction just before it lands:
The ball lands when $t = 0.904$ so $v_x = 5$ and $v_y = -8.86$.

The components of velocity are shown in figure 1.5.
The speed of the ball is

$$\sqrt{5^2 + 8.86^2} = 10.17\,\text{ms}^{-1}$$

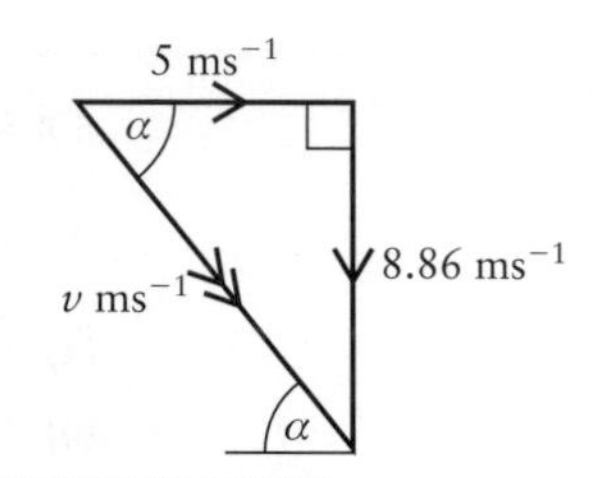

FIGURE 1.5

It hits the ground moving downwards at an angle α to the horizontal where

$$\tan \alpha = \frac{8.86}{5}$$
$$\alpha = 60.6°$$

Notice that the time forms a link between the motions in the two directions. You can often find the time from one equation and then substitute it in another to find out more information.

EXERCISE 1B

In this exercise take upwards as positive and use $9.8\,\text{ms}^{-2}$ for g in numerical questions.

1 In each case

(i) draw a diagram showing the initial velocity and path.

Write in component form

(ii) the velocity after time t s

(iii) the position after time t s.

(a) Initial position (0, 10 m); initial velocity $4\,\text{ms}^{-1}$ horizontally.

(b) Initial position (0, 7 m); initial velocity $10\,\text{ms}^{-1}$ at 35° above the horizontal.

(c) Initial position (0, 20 m); initial velocity $10\,\text{ms}^{-1}$ at 13° below the horizontal.

(d) Initial position O; initial velocity $7\,\text{m s}^{-1}$ horizontally and $24\,\text{m s}^{-1}$ vertically.

(e) Initial position (a, b) m; initial velocity $u_o\,\text{m s}^{-1}$ horizontally and $v_o\,\text{m s}^{-1}$ vertically.

2 In each case find

(i) the time taken for the projectile to reach its highest point

(ii) the maximum height above the origin.

(a) Initial position (0, 15 m); velocity $5\,\text{m s}^{-1}$ horizontally and $14.7\,\text{m s}^{-1}$ vertically.

(b) Initial position (0, 10 m); initial velocity $5\,\text{m s}^{-1}$ horizontally and $3\,\text{m s}^{-1}$ vertically.

3 Find the horizontal range for these projectiles which start from the origin.

(a) Initial velocity $2\,\text{m s}^{-1}$ horizontally and $7\,\text{m s}^{-1}$ vertically.

(b) Initial velocity $7\,\text{m s}^{-1}$ horizontally and $2\,\text{m s}^{-1}$ vertically.

(c) Sketch the paths of these two projectiles using the same axes.

FURTHER EXAMPLES

EXAMPLE 1.2

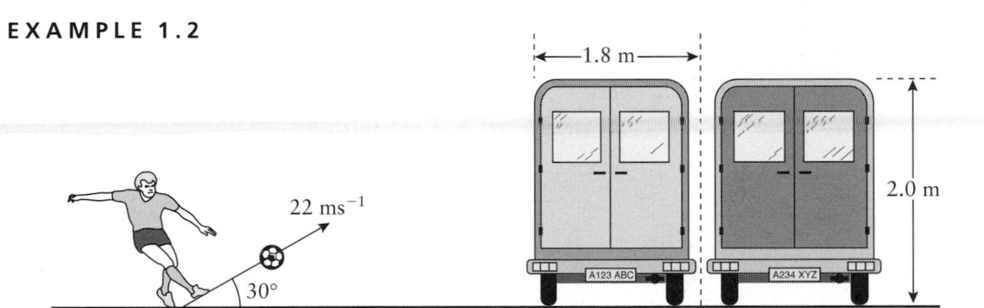

FIGURE 1.6

In this question use $10\,\text{m s}^{-2}$ for g and neglect air resistance.

In an attempt to raise money for a charity, participants are sponsored to kick a ball over some vans. The vans are each 2 m high and 1.8 m wide and stand on horizontal ground. One participant kicks the ball at an initial speed of $22\,\text{m s}^{-1}$ inclined at 30° to the horizontal.

(a) What are the initial values of the vertical and horizontal components of velocity?

(b) Show that while in flight the vertical height y metres at time t seconds satisfies the equation $y = 11t - 5t^2$.

Calculate at what times the ball is at least 2 m above the ground.

The ball should pass over as many vans as possible.

(c) Deduce that the ball should be placed about 3.8 m from the first van and find how many vans the ball will clear.

(d) What is the greatest vertical distance between the ball and the top of the vans?

[MEI]

Solution **(a)** *Initial velocity*

Horizontally $22\cos 30° = 19.05\,\text{ms}^{-1}$

Vertically $22\sin 30° = 11\,\text{ms}^{-1}$

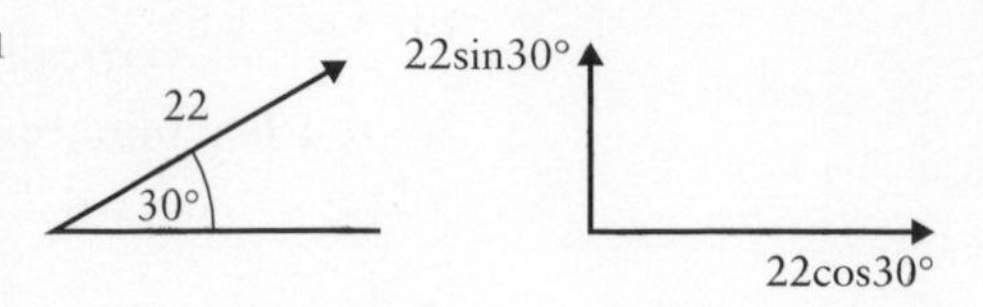

FIGURE 1.7

(b) *When the ball is above 2 m*

Using axes as shown and $s = ut + \frac{1}{2}at^2$ vertically

$$\Rightarrow \quad y = 11t - 5t^2$$

The ball is 2 m above the ground when $y = 2$, then

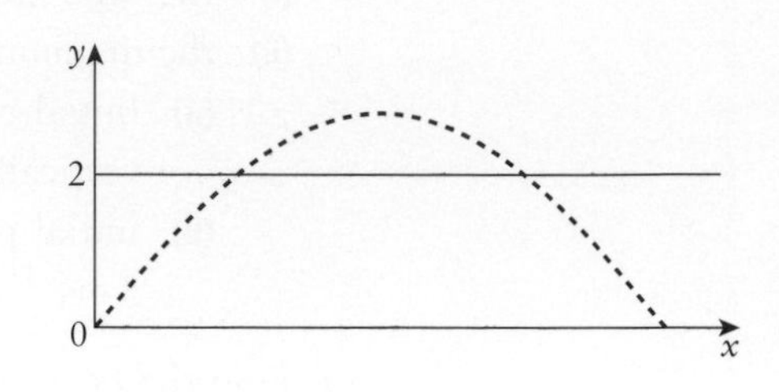

FIGURE 1.8

$$2 = 11t - 5t^2$$
$$5t^2 - 11t + 2 = 0$$
$$(5t - 1)(t - 2) = 0$$
$$t = 0.2 \text{ or } 2$$

$a = -10\ (\text{ms}^{-2})$ because the positive direction is upwards.

The ball is at least 2 m above the ground when $0.2 \leqslant t \leqslant 2$.

(c) *How many vans?*

Horizontally $s = ut + \frac{1}{2}at^2$ with $a = 0$

$$\Rightarrow \quad x = 19.05t$$

When $t = 0.2$ $\quad x = 3.81$ (at A)

When $t = 2$ $\quad x = 38.1$ (at B)

To clear as many vans as possible, the ball should be placed about 3.8 m in front of the first van.

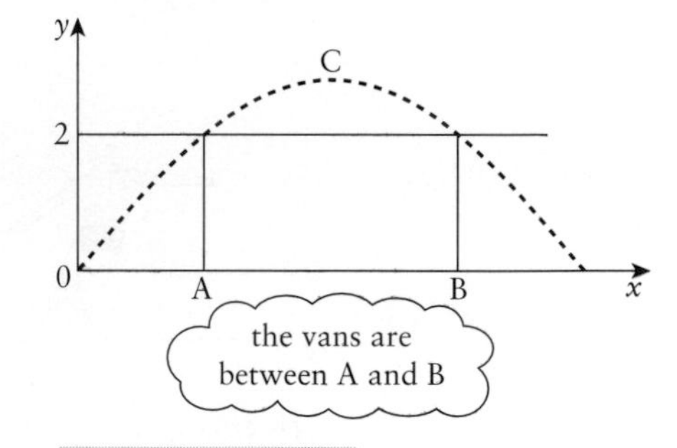

FIGURE 1.9

$$\text{AB} = 38.1 - 3.81\,\text{m} = 34.29\,\text{m}$$

$$\frac{34.29}{1.8} = 19.05$$

The maximum possible number of vans is 19.

(d) *Maximum height*

At the top (C), vertical velocity = 0, so using $v = u + at$ vertically

$$\Rightarrow \quad 0 = 11 - 10t$$
$$t = 1.1$$

Substituting in $y = 11t - 5t^2$, maximum height is $11 \times 1.1 - 5 \times 1.1^2 = 6.05\,\text{m}$

The ball clears the tops of the vans by about 4 m.

EXAMPLE 1.3

Sharon is diving into a swimming pool. During her flight she may be modelled as a particle. Her initial velocity is $1.8\,\text{ms}^{-1}$ at angle $30°$ above the horizontal and initial position 3.1 m above the water. Air resistance may be neglected.

(a) Find the greatest height above the water that Sharon reaches during her dive.

(b) Show that the time t, in seconds, that it takes Sharon to reach the water is given by $4.9t^2 - 0.9t - 3.1 = 0$ and solve this equation to find t.
Explain the significance of the other root to the equation.

Just as Sharon is diving a small boy jumps into the swimming pool. He hits the water at a point in line with the diving board and 1.5 m from its end.

(c) Is there an accident?

Solution

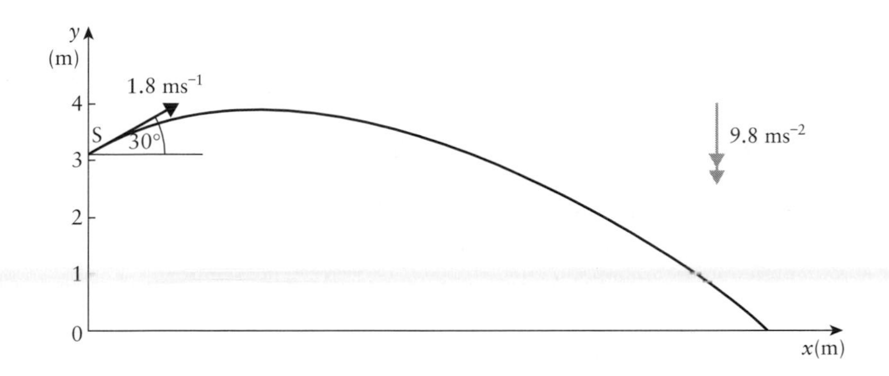

FIGURE 1.10

Referring to the axes shown

	Horizontal motion		Vertical motion	
initial position	0		3.1	
a	0		–9.8	
u	$u_x = 1.8\cos 30° = 1.56$		$u_y = 1.8\sin 30° = 0.9$	
v	$v_x = 1.56$	①	$v_y = 0.9 - 9.8t$	②
position	$x = 1.56t$	③	$y = 3.1 + 0.9t - 4.9t^2$	④

(a) At the top $v_y = 0$ $\quad 0 = 0.9 - 9.8t$ from ②

$t = 0.092$

When $t = 0.092$ $\quad y = 3.1 + 0.9 \times 0.092 - 4.9 \times 0.092^2$ from ④

$= 3.14$

Sharon's greatest height above the water is 3.14 m.

(b) Sharon reaches the water when $y = 0$

$$0 = 3.1 + 0.9t - 4.9t^2 \qquad \text{from ④}$$

$$4.9t^2 - 0.9t - 3.1 = 0$$

$$t = \frac{0.9 \pm \sqrt{0.9^2 + 4 \times 4.9 \times 3.1}}{9.8}$$

$$t = -0.71 \text{ or } 0.89$$

Sharon hits the water after 0.89 s. The negative value of t gives the point on the parabola at water level to the left of the point (S) where Sharon dives.

(c) At time t the horizontal distance from diving board,

$$x = 1.56t \qquad \text{from ③}$$

When Sharon hits the water $\quad x = 1.56 \times 0.89 = 1.39$

Assuming that the particles representing Sharon and the boy are located at their centres of mass, the difference of 11 cm between 1.39 m and 1.5 m is not sufficient to prevent an accident.

Note

When the point S is taken as the origin in the example above, the initial position is (0, 0) and $y = 0.9t - 4.9t^2$. In this case, Sharon hits the water when $y = -3.1$. This gives the same equation for t.

EXAMPLE 1.4

A boy kicks a small ball from the floor of a gymnasium with an initial velocity of $12\,\text{ms}^{-1}$ inclined at an angle α to the horizontal. Air resistance may be neglected.

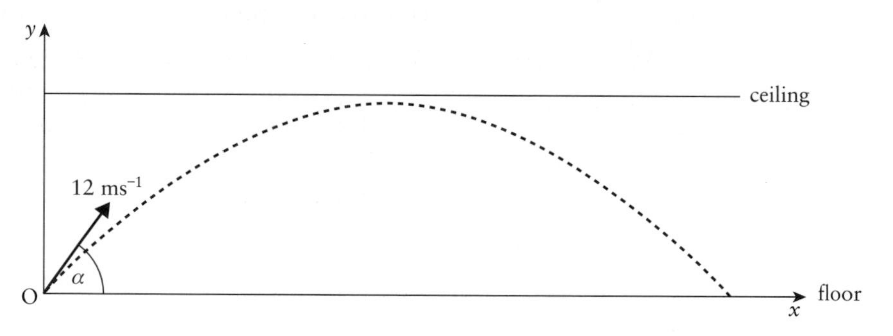

FIGURE 1.11

(a) Write down expressions in terms of α for the vertical speed of the ball and vertical height of the ball after t seconds.

The ball just fails to touch the ceiling which is 4 m high. The highest point of the motion of the ball is reached after T seconds.

(b) Use one of your expressions to show that $6\sin\alpha = 5T$ and the other to form a second equation involving $\sin\alpha$ and T.

(c) Eliminate $\sin\alpha$ from your two equations to show that T has a value of about 0.89.

(d) Find the horizontal range of the ball when kicked at $12\,\text{ms}^{-1}$ from the floor of the gymnasium so that it just misses the ceiling.

Use $10\,\text{ms}^{-2}$ for g in this question.

[MEI]

Solution

(a) *Vertical components*

speed $\quad v_y = 12\sin\alpha - 10t \quad$ ①

height $\quad y = (12\sin\alpha)t - 5t^2 \quad$ ②

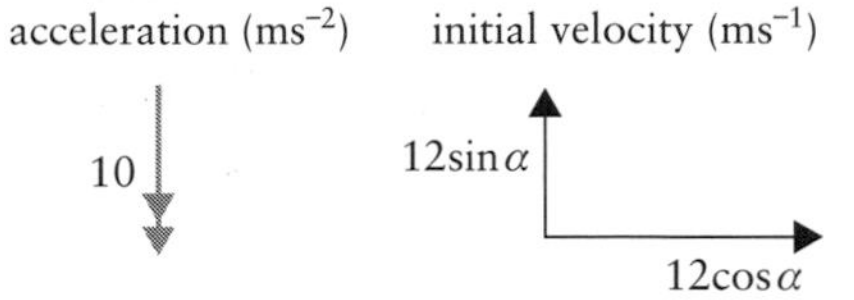

FIGURE 1.12

(b) *Time to highest point*

At the top $v_y = 0$ and $t = T$, so equation ① gives

$$12\sin\alpha - 10T = 0$$
$$12\sin\alpha = 10T$$
$$6\sin\alpha = 5T \qquad ③$$

When $t = T$, $y = 4$ so from ②

$$4 = (12\sin\alpha)T - 5T^2 \qquad ④$$

(c) Substituting for $6\sin\alpha$ from ③ into ④ gives

$$4 = 2 \times 5T \times T - 5T^2$$
$$4 = 5T^2$$
$$T = \sqrt{0.8} = 0.89$$

(d) *Range*

The path is symmetrical so the time of flight is $2T$ seconds.

Horizontally $a = 0$ and

$$u_x = 12\cos\alpha$$
$$\Rightarrow \quad x = (12\cos\alpha)t$$

The range is

$$12\cos\alpha \times 2T = 21.47\cos\alpha \text{ m}$$

From ③ $\quad 6\sin\alpha = 5T = 4.45$

$$\alpha = 47.87°$$

The range is $21.47\cos 47.87° = 14.4$ m.

Exercise 1C

Use 9.8 ms^{-2} for g in this exercise unless otherwise specified.

1 A ball is thrown from a point at ground level with velocity 20 ms^{-1} at 30° to the horizontal. The ground is level and horizontal and you should ignore air resistance. Take g to be 10 ms^{-2} and specify suitable axes.
   (a) Find the horizontal and vertical components of the ball's initial velocity.
   (b) Find the horizontal and vertical components of the ball's acceleration.
   (c) Find the horizontal distance travelled by the ball before its first bounce.
   (d) Find how long the ball takes to reach maximum height.
   (e) Find the maximum height reached by the ball.

2 Nick hits a golf ball with initial velocity 50 ms^{-1} at 35° to the horizontal.
   (a) Find the horizontal and vertical components of the ball's initial velocity.
   (b) Specify suitable axes and calculate the position of the ball at 1-second intervals for the first 6 seconds of its flight.
   (c) Draw a graph of the path of the ball (its trajectory) and use it to estimate
      (i) the maximum height of the ball
      (ii) the horizontal distance the ball travels before bouncing.
   (d) Calculate the maximum height the ball reaches and the horizontal distance it travels before bouncing. Compare your answers with the estimates you found from your graph.
   (e) State the modelling assumptions you made in answering this question.

3 Clare scoops a hockey ball off the ground, giving it an initial velocity of 19 ms^{-1} at 25° to the horizontal.
   (a) Find the horizontal and vertical components of the ball's initial velocity.
   (b) Find the time that elapses before the ball hits the ground.
   (c) Find the horizontal distance the ball travels before hitting the ground.
   (d) Find how long it takes for the ball to reach maximum height.
   (e) Find the maximum height reached.

   A member of the opposing team is standing 20 m away from Clare in the direction of the ball's flight.
   (f) How high is the ball when it passes her? Can she stop the ball?

4 A footballer is standing 30 m in front of the goal. He kicks the ball towards the goal with velocity 18 ms^{-1} and angle 55° to the horizontal. The height of the goal's crossbar is 2.5 m. Air resistance and spin may be ignored.
   (a) Find the horizontal and vertical components of the ball's initial velocity.
   (b) Find the time it takes for the ball to cross the goal-line.
   (c) Does the ball bounce in front of the goal, go straight into the goal or go over the crossbar?

   In fact the goalkeeper is standing 5 m in front of the goal and will stop the ball if its height is less than 2.8 m when it reaches him.
   (d) Does the goalkeeper stop the ball?

5 An aircraft is flying at a speed of $300\,\text{ms}^{-1}$ and maintaining an altitude of 10 000 m when a bolt becomes detached. Ignoring air resistance, find

(a) the time that the bolt takes to reach the ground

(b) the horizontal distance between the point where the bolt leaves the aircraft and the point where it hits the ground

(c) the speed of the bolt when it hits the ground

(d) the angle to the horizontal at which the bolt hits the ground.

6 Reena is learning to serve in tennis. She hits the ball from a height of 2 m. For her serve to be legal it must pass over the net, which is 12 m away from her and 0.91 m high, and it must land within 6.4 m of the net.
Make the following modelling assumptions to answer the questions.

- She hits the ball horizontally.
- Air resistance may be ignored.
- The ball may be treated as a particle.
- The ball does not spin.
- She hits the ball straight down the middle of the court.

(a) How long does the ball take to fall to the level of the top of the net?

(b) How long does the ball take from being hit to first reaching the ground?

(c) What is the lowest speed with which Reena must hit the ball to clear the net?

(d) What is the greatest speed with which she may hit it if it is to land within 6.4 m of the net?

7 A stunt motorcycle rider attempts to jump over a gorge 50 m wide. He uses a ramp at 25° to the horizontal for his take-off and has a speed of $30\,\text{ms}^{-1}$ at this time.

(a) Assuming that air resistance is negligible, find out whether the rider crosses the gorge successfully.

The stunt man actually believes that in any jump the effect of air resistance is to reduce his distance by 40%.

(b) Calculate his minimum safe take-off speed for this jump.

8 To kick a goal in rugby you must kick the ball over the crossbar of the goal posts (height 3.0 m), between the two uprights. Dafydd Evans attempts a kick from a distance of 35 m. The initial velocity of the ball is $20\,\text{ms}^{-1}$ at 30° to the horizontal. The ball is aimed between the uprights and no spin is applied.

(a) How long does it take the ball to reach the goal posts?

(b) Does it go over the crossbar?

Later in the game, Dafydd takes another kick from the same position and hits the crossbar.

(c) Given that the initial velocity of the ball in this kick was also at 30° to the horizontal, find the initial speed.

Many rugby kickers choose to give the ball spin.

(d) What effect does spin have upon the flight of the ball?

9 In this question take g to be $10\,\text{ms}^{-2}$. A catapult projects a small pellet at speed $20\,\text{ms}^{-1}$ and can be directed at any angle to the horizontal.

(a) Find the range of the catapult when the angle of projection is
(i) $30°$ **(ii)** $40°$ **(iii)** $45°$ **(iv)** $50°$ **(v)** $60°$.

(b) Show algebraically that the range is the same when the angle of projection is α as it is when the angle is $90 - \alpha$.

The catapult is angled with the intention that the pellet should hit a point on the ground 36 m away.

(c) Verify that one appropriate angle of projection would be $32.1°$ and write down another suitable angle.

10 A cricketer hits the ball on the half-volley, that is when the ball is at ground level. The ball leaves the ground at an angle of $30°$ to the horizontal and travels towards a fielder standing on the boundary 60 m away.

(a) Find the initial speed of the ball if it hits the ground for the first time at the fielder's feet.

(b) Find the initial speed of the ball if it is at a height of 3.2 m (well outside the fielder's reach) when it passes over the fielder's head.

In fact the fielder is able to catch the ball without moving provided that its height, h m, when it reaches him satisfies the inequality $0.25 \leqslant h \leqslant 2.1$.

(c) Find a corresponding range of values for u, the initial speed of the ball.

11 A horizontal tunnel has a height of 3 m. A ball is thrown inside the tunnel with an initial speed of $18\,\text{ms}^{-1}$. What is the greatest horizontal distance that the ball can travel before it bounces for the first time?

THE PATH OF A PROJECTILE

Look at the equations

$$x = 20t$$
$$y = 6 + 30t - 5t^2$$

They represent the path of a projectile.

These equations give x and y in terms of a third variable t. (They are called *parametric equations* and t is the *parameter*.)

You can find the *cartesian equation* connecting x and y directly by eliminating t as follows

$$x = 20t \Rightarrow t = \frac{x}{20}$$

so $$y = 6 + 30t - 5t^2$$

can be written as $y = 6 + 30 \times \frac{x}{20} - 5 \times \left(\frac{x}{20}\right)^2$

$$y = 6 + 1.5x - \frac{x^2}{80}$$

EXERCISE 1D

In this exercise make the simplification that g is 10 ms⁻².

1 Find the cartesian equation of the path of these projectiles by eliminating the parameter t.

(a) $x = 4t$ $\quad y = 5t^2$

(b) $x = 5t$ $\quad y = 6 + 2t - 5t^2$

(c) $x = 2 - t$ $\quad y = 3t - 5t^2$

(d) $x = 1 + 5t$ $\quad y = 8 + 10t - 5t^2$

(e) $x = ut$ $\quad y = 2ut - \frac{1}{2}gt^2$

2 A particle is projected with initial velocity $50\,\text{ms}^{-1}$ at an angle of 36.9° to the horizontal. The point of projection is taken to be the origin, with the x axis horizontal and the y axis vertical in the plane of the particle's motion.

(a) Show that at time t s, the height of the particle in metres is given by $y = 30t - 5t^2$ and write down the corresponding expression for x.

(b) Eliminate t between your equations for x and y to show that

$$y = \frac{3x}{4} - \frac{x^2}{320}.$$

(c) Plot the graph of y against x using a scale of 2 cm for 10 m along both axes.

(d) Mark on your graph the points corresponding to the position of the particle after 1, 2, 3, 4, ... seconds.

3 A golfer hits a ball with initial velocity $50\,\text{ms}^{-1}$ at an angle α to the horizontal where $\sin\alpha = 0.6$.

(a) Find the equation of its trajectory, assuming that air resistance may be neglected.

The flight of the ball is recorded on film and its position vector, from the point where it was hit, is calculated. The x and y axes are horizontal and vertical in the plane of the ball's motion. The results (to the nearest 0.5 m) are as follows:

Time (s)	Position (m)	Time (s)	Position (m)
0	(0, 0)	4	(152, 39)
1	(39.5, 24.5)	5	(187.5, 24.5)
2	(78, 39)	6	(222, 0)
3	(116.5, 44)		

(b) On the same piece of graph paper draw the trajectory you found in part (a) and that found from analysing the film. Compare the two graphs and suggest a reason for any differences.

(c) It is suggested that the horizontal component of the resistance to the motion of the golf ball is almost constant. Are the figures consistent with this?

ACCESSIBLE POINTS

The equation of the path of a projectile can be used to decide whether certain points can be reached by the projectile. The next two examples illustrate how this can be done.

EXAMPLE 1.5

A projectile is launched from the origin with an initial velocity $20\,\text{ms}^{-1}$ at an angle of $30°$ to the horizontal.

(a) Write down the position of the projectile after time t.

(b) Show that the equation of the path is the parabola $y = 0.578x - 0.016x^2$.

(c) Find y when $x = 3$.

(d) Decide whether the projectile can hit a point 6 m above the ground.

Solution

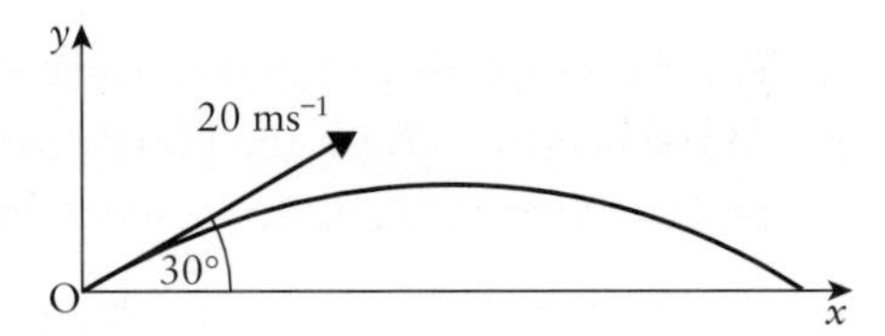

FIGURE 1.13

(a) Using horizontal and vertical components for the position

Horizontally $x = (20\cos 30°)t$

Vertically $y = (20\sin 30°)t - 4.9t^2$

$\Rightarrow$ $x = 17.3t$ ①

and $y = 10t - 4.9t^2$ ②

(b) From equation ① $t = \dfrac{x}{17.3}$

Substituting this into equation ② for y gives

$$y = 10 \times \frac{x}{17.3} - 4.9 \times \frac{x^2}{17.3^2}$$

$$\Rightarrow \quad y = 0.578x - 0.016x^2$$

(c) When $x = 3$ $\qquad y = 0.578 \times 3 - 0.016 \times 9 = 1.59$

(d) When $y = 6$ $\qquad 6 = 0.578x - 0.016x^2$

$$\Rightarrow \qquad 0.016x^2 - 0.578x + 6 = 0$$

In this quadratic equation the discriminant
$b^2 - 4ac = 0.578^2 - 4 \times 0.016 \times 6 = -0.0499$

You cannot find the square root of this negative number, so the equation cannot be solved for x. The projectile cannot hit a point 6 m above the ground.

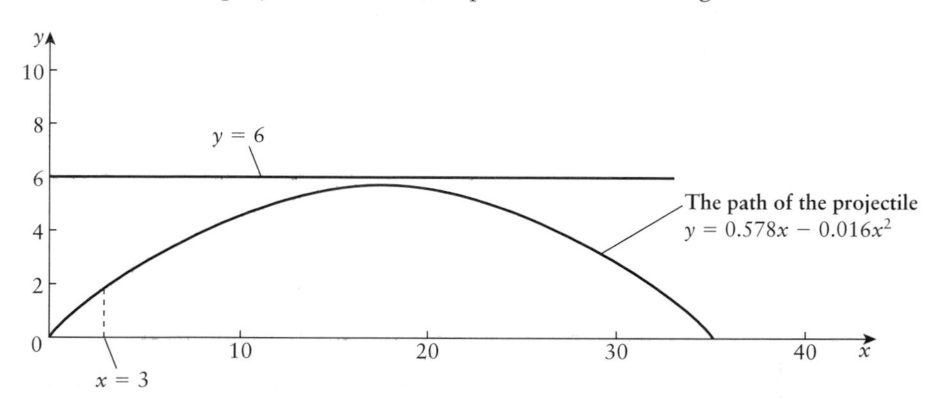

FIGURE 1.14

Looking at the sign of the discriminant of a quadratic equation is a good way of deciding whether points are within the range of a projectile. The example above is a quadratic in x. Example 1.6 involves a quadratic equation in $\tan\alpha$ but the same idea is used in part **(c)** to determine whether certain points are within range.

EXAMPLE 1.6

A ball is hit from the origin with a speed of $14\,\text{m}\,\text{s}^{-1}$ at an angle α to the horizontal.

(a) Find the equation of the path of the ball in terms of $\tan\alpha$.

(b) Find the two values of α which ensure that the ball passes through the point (5, 2.5).

(c) Decide whether the ball can pass through the points
(i) (10, 7.5) **(ii)** (8, 9).

Solution

(a) The components of the initial velocity are $14\cos\alpha$ and $14\sin\alpha$, so the path of the ball is given by the equations

$$x = (14\cos\alpha)t \text{ and } y = (14\sin\alpha)t - 4.9t^2$$

$$\Rightarrow \qquad t = \frac{x}{14\cos\alpha} \text{ and } y = 14\sin\alpha \times \frac{x}{14\cos\alpha} - 4.9 \times \frac{x^2}{(14\cos\alpha)^2}$$

$$\Rightarrow \qquad y = x\tan\alpha - \frac{x^2}{40\cos^2\alpha}$$

You can then express this in a more useful form by using two trigonometrical identities

$$\frac{1}{\cos\alpha} = \sec\alpha \text{ and } \sec^2\alpha = 1 + \tan^2\alpha$$

So the equation of the path is $y = x\tan\alpha - \frac{1}{40}x^2(1 + \tan^2\alpha)$ ①

(ii) When $x = 5$ and $y = 2.5$ $\quad 2.5 = 5\tan\alpha - \frac{25}{40}(1 + \tan^2\alpha)$

multiply by $\frac{40}{25}$ → $\quad 4 = 8\tan\alpha - (1 + \tan^2\alpha)$

$$\tan^2\alpha - 8\tan\alpha + 5 = 0$$

This is a quadratic in $\tan\alpha$ which has solution $\tan\alpha = \dfrac{8 \pm \sqrt{64 - 20}}{2}$.

So $\tan\alpha = 7.32$ or 0.68 giving possible angles of projection of 82° and 34°.

(c) (i) For the point (10, 7.5), equation ① gives

$$7.5 = 10\tan\alpha - \frac{10^2}{40}(1 + \tan^2\alpha)$$

$$7.5 = 10\tan\alpha - 2.5(1 + \tan^2\alpha)$$

$$\tan^2\alpha - 4\tan\alpha + 4 = 0 \leftarrow \text{divide by 2.5}$$

The discriminant $b^2 - 4ac = 16 - 16 = 0$

This zero discriminant means that the equation has two equal roots, $\tan\alpha = 2$. The point (10, 7.5) lies on the path of the projectile but there is only one possible angle of projection.

(ii) For the point (8, 9), equation ① gives $9 = 8\tan\alpha - \frac{8^2}{40}(1 + \tan^2\alpha)$

$$1.6\tan^2\alpha - 8\tan\alpha + 10.6 = 0$$

The discriminant is $64 - 4 \times 1.6 \times 10.6 = -3.84$ and this has no real square root.

This equation cannot be solved to find $\tan\alpha$ so the projectile cannot pass through the point (8, 9) if it has an initial speed of $14\,\text{ms}^{-1}$.

The diagram shows all the results.

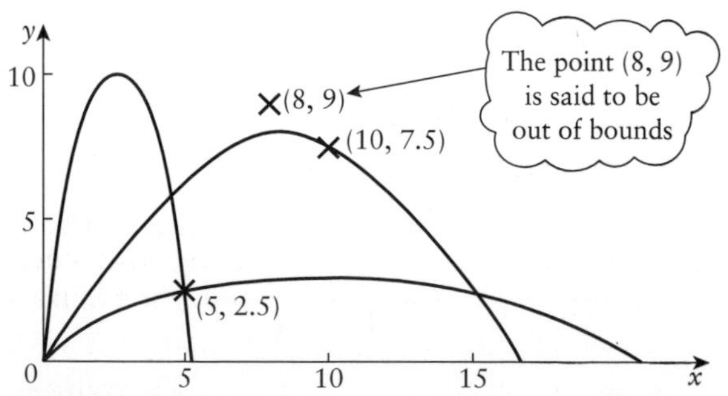

FIGURE 1.15

EXERCISE 1E

Use 10 ms^{-2} for g in this exercise unless otherwise instructed and use the modelling assumptions that air resistance can be ignored and the ground is horizontal.

1 A projectile is launched from the origin with an initial velocity 30 ms^{-1} at an angle of 45° to the horizontal.

(a) Write down the position of the projectile after time t.

(b) Show that the equation of the path is the parabola $y = x - 0.011x^2$.

(c) Find y when $x = 10$.

(d) Find x when $y = 20$.

2 Jack throws a cricket ball at a wicket 0.7 m high with velocity 10 ms^{-1} at 14° above the horizontal. The ball leaves his hand 1.5 m above the origin.

(a) Show that the equation of the path is the parabola
$y = 1.5 + 0.25x - 0.053x^2$.

(b) How far from the wicket is he standing if the ball just hits the top?

3 While practising his tennis serve, Matthew hits the ball from a height of 2.5 m with a velocity of magnitude 25 ms^{-1} at an angle of 5° above the horizontal as shown in the diagram. In this question, take $g = 9.8$ ms^{-2}.

(a) Show that while in flight
$y = 2.5 + 0.087x - 0.0079x^2$.

(b) Find the horizontal distance from the serving point to the spot where the ball lands.

(c) Determine whether the ball would clear the net, which is 1 m high and 12 m from the serving position in the horizontal direction.

4 Ching is playing volleyball. She hits the ball with initial speed u ms^{-1} from a height of 1 m at an angle of 35° to the horizontal.

(a) Define a suitable origin and x and y axes and find the equation of the trajectory of the ball in terms of x, y and u.

The rules of the game require the ball to pass over the net, which is at height 2 m, and land inside the court on the other side, which is of length 5 m. Ching hits the ball straight along the court and is 3 m from the net when she does so.

(b) Find the minimum value of u for the ball to pass over the net.

(c) Find the maximum value of u for the ball to land inside the court.

5 A ball is hit from the origin with a speed of $10\,\text{ms}^{-1}$ at an angle α to the horizontal.

(a) Show that the equation of the path is $y = x\tan\alpha - 0.05x^2(1 + \tan^2\alpha)$.

(b) Find the two values of α which ensure that the ball passes through the point (2, 3).

(c) Decide whether the ball can pass through the points

(i) (5, 3) **(ii)** (3, 5).

6 A ball is hit from the point (0, 2) with a speed of $40\,\text{ms}^{-1}$ at an angle α to the horizontal.

(a) Find the equation of the path of the ball in terms of $\tan\alpha$.

(b) Find the two values of α which ensure that the ball passes through the point (18, 0).

(c) Can the ball pass through the points **(i)** (160, 5) **(ii)** (150, 8)?

7 Tennis balls are delivered from a machine at a height of 1 m and with a speed of $25\,\text{ms}^{-1}$. Dan stands 15 m from the machine and can reach to a height of 2.5 m.

(a) Write down the equations for the horizontal and vertical displacements x and y after time t seconds for a ball which is delivered at an angle $\alpha°$ to the horizontal.

(b) Eliminate t from your equations to obtain the equation for y in terms of x and $\tan\alpha$.

(c) Can Dan reach a ball when $\tan\alpha$ is

(i) 0.5 **(ii)** 0.2?

(d) Determine the values of α for which Dan can just hit the ball.

8 A high pressure hose is used to water a horizontal garden. The jet of water is modelled as a stream of small droplets (i.e. particles) projected at $15\,\text{ms}^{-1}$ at an angle α to the horizontal from a point 1.2 m above the garden. Air resistance may be neglected. The diagram shows this situation as well as the origin O and the axes. The unit of each axis is the metre.

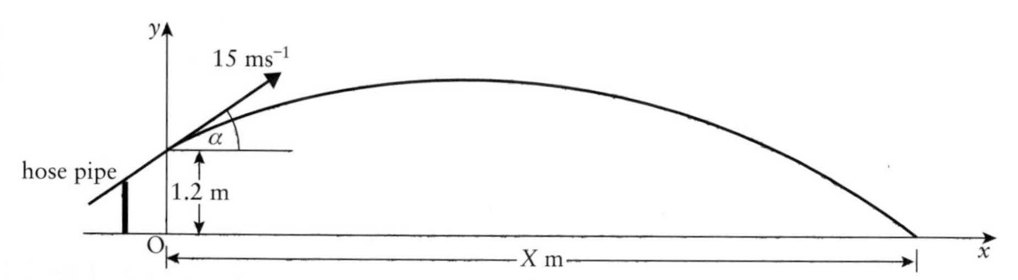

(a) Show that the vertical height of a droplet at time t is given by

$$y = 1.2 + 15t\sin\alpha - 5t^2.$$

Write down the corresponding equation for the horizontal distance.

The horizontal range of a droplet projected at angle α is X m, as shown in the diagram.

(b) Use your answers from part **(a)** to deduce that

$$X^2\tan^2\alpha - 45X\tan\alpha + X^2 - 54 = 0.$$

(c) Show that it is possible to adjust the angle of projection so that the water lands on the point (20, 0) but that the point (30, 0) is out of range.

(d) Does the maximum value of X occur when $\alpha = 45°$? Explain your answer briefly.

9 A golf ball is driven from the tee with speed $30\sqrt{2}$ ms^{-1} at an angle α to the horizontal.

(a) Show that during its flight the horizontal and vertical displacements x and y of the ball from the tee satisfy the equation

$$y = x\tan\alpha - \frac{x^2}{360}(1 + \tan^2\alpha).$$

(b) The golf ball just clears a tree 5 m high which is 150 m horizontally from the tee. Find the two possible values of $\tan\alpha$.

(c) Use the discriminant of the quadratic equation in $\tan\alpha$ to find the greatest distance by which the golf ball can clear the tree and find the value of $\tan\alpha$ in this case.

(d) The ball is aimed at the hole which is on the green immediately behind the tree. The hole is 160 m from the tee. What is the greatest height the tree could be without making it impossible to hit a hole in one? (Hint: $2\sin\alpha\cos\alpha = \sin 2\alpha$)

10 A particle is projected up a slope of angle β where $\tan\beta = \frac{1}{2}$. The initial velocity of the particle is 50 ms^{-1} at angle α to the horizontal where $\sin\alpha = \frac{4}{5}$. The x and y axes are taken from the point of projection and are in the plane of the particle's motion as shown below.

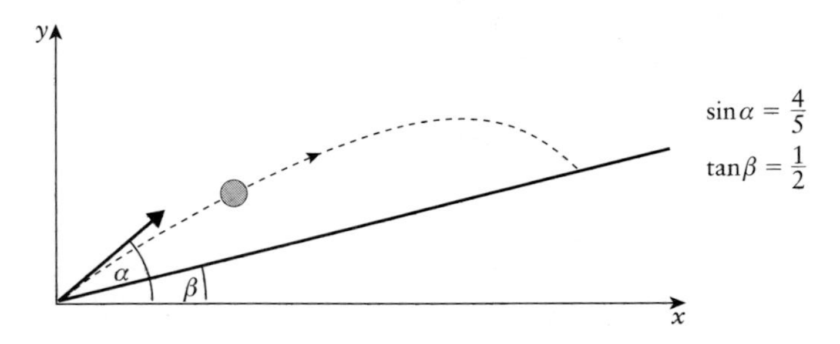

(a) Use values of g and $\tan\alpha$ to find the equation of the trajectory of the particle.

(b) Explain why the equation of the slope is $y = \frac{1}{2}x$.

(c) Solve the equations of the trajectory and the slope to find the coordinates of the point where the particle hits the slope.

(d) What is the range of the particle up the slope?

Another particle is then projected from the same point and with the same initial speed but at an angle of 45° to the horizontal.

(e) Find the range of this particle up the slope.

(f) Does an angle of projection of 45° to the horizontal result in the particle travelling the maximum distance up the slope?

General equations

The work done in this chapter can now be repeated for the general case using algebra. Assume a particle is projected from the origin with speed u at an angle α to the horizontal and that the only force acting on the particle is the force due to gravity. The x and y axes are horizontal and vertical through the origin, O, in the plane of motion of the particle.

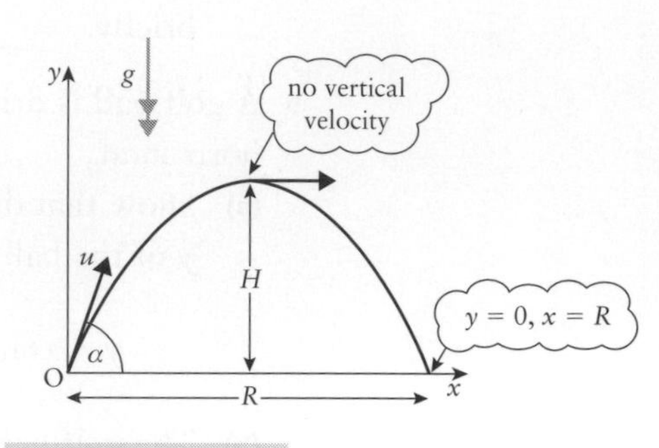

Figure 1.16

The components of velocity and position

	Horizontal motion		Vertical motion	
Initial position	0		0	
a	0		$-g$	
u	$u_x = u\cos\alpha$		$u_y = u\sin\alpha$	
v	$v_x = u\cos\alpha$	①	$v_y = u\sin\alpha - gt$	②
position	$x = ut\cos\alpha$	③	$y = ut\sin\alpha - \frac{1}{2}gt^2$	④

$ut\cos\alpha$ is preferable to $u\cos\alpha t$ because this could mean $u\cos(\alpha t)$, which is incorrect

The maximum height

At its greatest height, the vertical component of velocity is zero.
From equation ②

$$u\sin\alpha - gt = 0$$

$$t = \frac{u\sin\alpha}{g}$$

Substitute in equation ④ to obtain the height of the projectile

$$y = u \times \frac{u\sin\alpha}{g} \times \sin\alpha - \tfrac{1}{2}g \times \frac{(u\sin\alpha)^2}{g^2}$$

$$= \frac{u^2\sin^2\alpha}{g} - \frac{u^2\sin^2\alpha}{2g}$$

The greatest height is $H = \dfrac{u^2\sin^2\alpha}{2g}$

THE TIME OF FLIGHT

When the projectile hits the ground, $y = 0$.

From equation ④ $\quad y = ut\sin\alpha - \frac{1}{2}gt^2$

$$0 = ut\sin\alpha - \tfrac{1}{2}gt^2$$

$$0 = t(u\sin\alpha - \tfrac{1}{2}gt)$$

The solution $t = 0$ is at the start of the motion

The time of flight is $\quad t = \dfrac{2u\sin\alpha}{g}$

THE RANGE

The range of the projectile is the value of x when $t = \dfrac{2u\sin\alpha}{g}$

From equation ③ $\quad x = ut\cos\alpha$

$$\Rightarrow \quad R = u \times \frac{2u\sin\alpha}{g} \times \cos\alpha$$

$$R = \frac{2u^2\sin\alpha\cos\alpha}{g}$$

It can be shown that $2\sin\alpha\cos\alpha = \sin 2\alpha$, so the range can be expressed as

$$R = \frac{u^2\sin 2\alpha}{g}$$

The range is a maximum when $\sin 2\alpha = 1$, that is when $2\alpha = 90°$ or $\alpha = 45°$. The maximum possible horizontal range for projectiles with initial speed u is

$$R_{max} = \frac{u^2}{g}$$

THE EQUATION OF THE PATH

From equation ③ $\quad t = \dfrac{x}{u\cos\alpha}$

Substitute into equation ④ to give

$$y = u \times \frac{x}{u\cos\alpha} \times \sin\alpha - \tfrac{1}{2}g \times \frac{x^2}{(u\cos\alpha)^2}$$

$$y = x\frac{\sin\alpha}{\cos\alpha} - \frac{gx^2}{2u^2\cos^2\alpha}$$

$$y = x\tan\alpha - \frac{gx^2}{2u^2}(1 + \tan^2\alpha)$$

It is important that you understand the methods used to derive these formulae and don't rely on learning the results off by heart. They are only true when the given assumptions apply and the variables are as defined in figure 1.16.

Exercise 1F *Examination-style questions*

1 A ball rolls off a horizontal table of height 0.8 m and lands a horizontal distance of 1.5 m away.
  (a) Find the time it takes to reach the floor.
  (b) Find the speed with which it left the table.
  (c) Find, in degrees to 1 decimal place, the angle to the horizontal at which the ball is moving just before it hits the ground.

2 A golf ball is hit with velocity 50 ms^{-1} at 30° above the horizontal. The motion of the ball is modelled as a particle moving freely under gravity over horizontal ground.
  (a) Find the distance from the point at which it is struck to the point at which it lands.
  (b) Find the greatest height reached during the motion.
  (c) A tree which is 12 m tall is at a horizontal distance of 25 m from the initial position along the line of flight of the ball. Does the ball clear the tree?

3 In a first model of a serve at tennis, the ball is treated as a particle which is projected horizontally at a height of 2.5 m above the horizontal ground. Air resistance is ignored.
  (a) The ball must pass over the net, which is of height 0.91 m at a horizontal distance of 11.89 m from the server. Find the minimum velocity with which the ball must be served.
  (b) The ball must hit the ground within a horizontal distance of 18.29 m from the server. Find the maximum velocity with which the ball must be served.
  (c) State, briefly, one way in which the model can be made more realistic.

4 A firework rocket burns out when it is at a height of 60 m and moving with velocity 60 ms^{-1} at an angle α above the horizontal, where $\tan\alpha = \frac{4}{3}$. The empty case may then be modelled as a particle moving freely under gravity, and the ground as a horizontal plane. Find
  (a) the time it takes to fall to the ground
  (b) the horizontal distance between the point at which it burned out and the landing point
  (c) the speed with which it lands
  (d) the angle to the horizontal at which it is travelling just before it lands.

5 A football is kicked with velocity 20 ms^{-1} straight down the middle of a long horizontal corridor. The corridor is of height 3 m.
  By modelling the football as a particle and ignoring air resistance, find
  (a) the greatest angle, in degrees to 1 decimal place, above the horizontal at which the ball can be kicked and not hit the ceiling

(b) the time that the ball is in the air when the ball is kicked at that angle

(c) the horizontal distance travelled before the ball hits the ground in that case.

A teacher of height 1.73 m is standing 6 m away from the point at which the ball was kicked.

(d) Does the ball miss the teacher?

6 When taking a penalty kick in football, a player selects a target just in the goal at a horizontal distance of 11.5 m from the penalty spot and at a height of 2.3 m. The player kicks the ball so that this is the highest point of its path. Find

(a) the vertical component of the ball's initial velocity

(b) the time taken for the ball to reach the goal

(c) the horizontal component of the ball's initial velocity

(d) the angle, in degrees to 1 decimal place, above the horizontal at which the ball is kicked.

7 Juliet is on her balcony and throws a small object to Romeo on the horizontal ground below. She throws it with a velocity of $5\,\text{ms}^{-1}$ at an angle α below the horizontal, where $\tan\alpha = \frac{3}{4}$. Romeo catches it at a vertical distance of 3 m below the point of projection. Find

(a) the horizontal distance between Romeo and Juliet

(b) the speed with which the object is moving as Romeo catches it

(c) the angle to the horizontal, in degrees to 1 decimal place, at which the object is moving just before Romeo catches it.

8 A cat is trying to jump up on top of a wardrobe. The wardrobe is of height 2.2 m with a horizontal top, and stands on a horizontal floor in a room whose ceiling is parallel to the floor and 2.5 m above it. The cat is standing on the floor 1 m away from the wardrobe.

(a) Find the minimum vertical component of the cat's initial velocity if it is to reach the height of the wardrobe.

(b) Find the maximum vertical component of the cat's initial velocity if it is not to hit the ceiling.

The cat actually uses this maximum value.

(c) Find the times between which it is above the height of the wardrobe.

(d) Hence find a range of values for the horizontal component of the cat's initial velocity.

9 A cricket ball, which may be modelled as a particle moving freely under gravity, is struck from a height of 0.5 m above a horizontal field with a velocity of $26\,\text{ms}^{-1}$ at an angle α above the horizontal, where $\tan\alpha = \frac{5}{12}$.

(a) Between what times is the ball out of reach of the fielders, who can reach 2.5 m above the field?

(b) The fielders wish to catch the ball before it reaches the ground. How far from the bat should the fielders be placed in order to do this?

10 A particle is projected with speed u at an angle of α above the horizontal, and moves freely under gravity alone.

(a) Show that the particle returns to the height at which it was launched after it has travelled a horizontal distance of

$$\frac{u^2 \sin 2\alpha}{g}.$$

On a shooting range, the firing point and the target are at the same height. The range of a rifle is adjusted by altering the angle to the horizontal at which the bullet is projected. The bullet may be modelled as a particle moving freely under gravity with an initial speed of $400\,\text{ms}^{-1}$.

(b) Find the angle, in degrees to 3 significant figures, at which the bullet should be fired to give the rifle a range of 500 m.

(c) A shot is aimed correctly at a target which is 100 m away, but the sights have been left set at a range of 500 m. Find the height above the point of aim at which the target is hit.

11 An aircraft is on a training exercise to practise precision bombing. It flies in a straight line with a constant speed of $300\,\text{ms}^{-1}$ at a constant height of 1960 m above horizontal ground. It releases a bomb which then moves freely under gravity.

(a) Find the time taken for the bomb to reach the ground.

(b) Find, in degrees to 1 decimal place, the angle to the horizontal at which the bomb is moving immediately before it hits the ground.

The target for the bomb is a point on the ground directly ahead of the aircraft. An instrument on the aircraft measures, at any instant, the angle θ which the line between the aircraft and the target makes with the horizontal.

(c) Find, in degrees to 1 decimal place, the value of θ at the instant when the bomb should be released in order to hit the target.

[Edexcel]

12

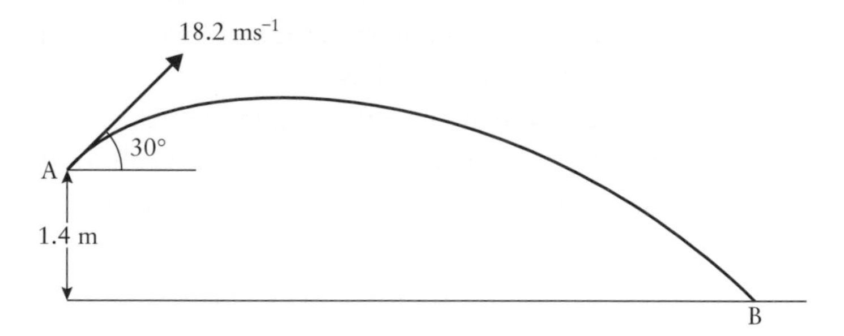

A batsman hits a cricket ball, giving it a speed of $18.2\,\text{ms}^{-1}$ at an angle of elevation of 30°. The point A where he hits the ball is 1.4 m above the ground, which is horizontal. The ball then moves freely under gravity and hits the ground at the point B, as shown in the diagram. By modelling the ball as a particle and ignoring air resistance, find

(a) the time taken by the ball to travel from A to B,

(b) the horizontal distance of B from A, giving your answer in metres to 1 decimal place.

[Edexcel]

13 A particle is projected with speed u at an angle α *below* the horizontal and moves freely under gravity alone. When the particle has moved a horizontal distance x, it has moved a vertical distance y.

(a) Show that

$$y = x\tan\alpha + \frac{gx^2}{2u^2\cos\alpha}$$

An archer sits on top of a vertical tower which stands on level ground. He sees a rabbit sitting at a point on the ground which is a horizontal distance of 20 m from the base of the tower. The archer fails to allow for any effect of gravity and aims an arrow directly at the point on the ground where the rabbit is sitting. The height of the arrow is initially 10 m above the ground, and the initial speed of the arrow is 28 ms^{-1}. By modelling the arrow as a particle moving freely under gravity alone,

(b) find the distance by which the arrow falls short of the rabbit.

[Edexcel]

14 A cricket ball is hit from a height of 0.9 m above horizontal ground with a speed of 25 ms^{-1} at an angle α above the horizontal, where $\tan\alpha = \frac{7}{24}$. The motion of the ball is modelled as that of a particle moving freely under gravity.

(a) Find the length of time for which the ball is at least 3 m above the ground.

The ball is caught by a fielder who is 32 m horizontally from the point where it was struck.

(b) Find the height above the ground at which the fielder makes the catch.

(c) State, briefly, one way in which the above model may be refined in order to make it more realistic.

[Edexcel]

KEY POINTS

1 Modelling assumptions for projectile motion

- a projectile is a particle
- it is not powered
- the air has no effect on its motion.

2 Projectile motion is usually considered in terms of horizontal and vertical components.

When the initial position is at O

Angle of projection = α

Initial velocity x component = $u\cos\alpha$
y component = $u\sin\alpha$

Acceleration x component = 0
y component = $-g$

At time t, velocity $v = u + at$ in both directions

$v_x = u\cos\alpha + 0t$

$v_x = u\cos\alpha$ ①

$v_y = u\sin\alpha - gt$ ②

Displacement $s = ut + \frac{1}{2}at^2$ in both directions

$x = u\cos\alpha t + \frac{1}{2}0t^2$

$x = ut\cos\alpha$ ③

$y = u\sin\alpha t + \frac{1}{2}(-g)t^2$

$y = ut\sin\alpha - \frac{1}{2}gt^2$ ④

3 At maximum height $v_y = 0$.

4 $y = 0$ when the projectile lands.

5 The time to hit the ground is twice the time to maximum height.

6 The equation of the path of a projectile is $y = x\tan\alpha - \frac{gx^2}{2u^2}(1 + \tan^2\alpha)$.

7 For given u, x and y, this equation can be written as a quadratic in $\tan\alpha$.

When there are

- 2 roots the point is within bounds
- 1 root the point is at the limit
- 0 roots the point is out of bounds.

Chapter two

CALCULUS METHODS

They are neither finite quantities, or quantities infinitely small,
nor yet nothing. May we not call them the ghosts of departed quantities?

The Analyst, ***George Berkeley***

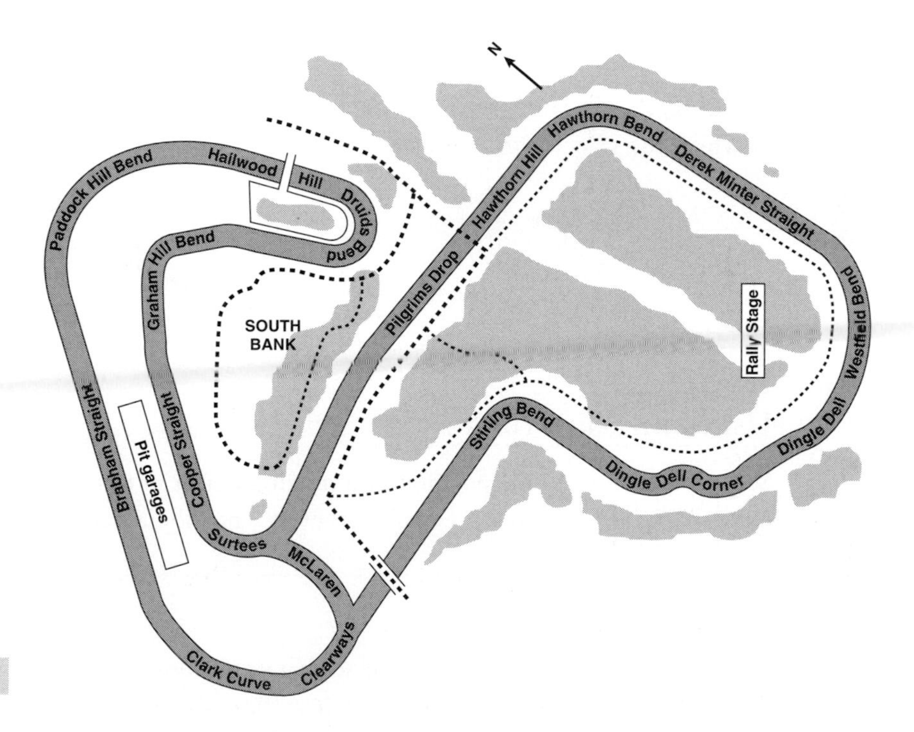

FIGURE 2.1

So far you have studied motion with constant acceleration in a straight line, but the motion of a car round the Brand's Hatch racing circuit shown in the diagram is much more complex. In this chapter you will see how to deal first with variable acceleration and later with motion in two dimensions.

MOTION IN ONE DIMENSION

The equations you have used for constant acceleration do not apply when the acceleration varies. You need to go back to first principles.

Consider how displacement, velocity and acceleration are related to each other. The velocity of an object is the rate at which its position changes with time. When the velocity is not constant the position–time graph is a curve.

The rate of change of the position is the gradient of the tangent to the curve. You can find this by differentiating.

$$v = \frac{ds}{dt} \quad ①$$

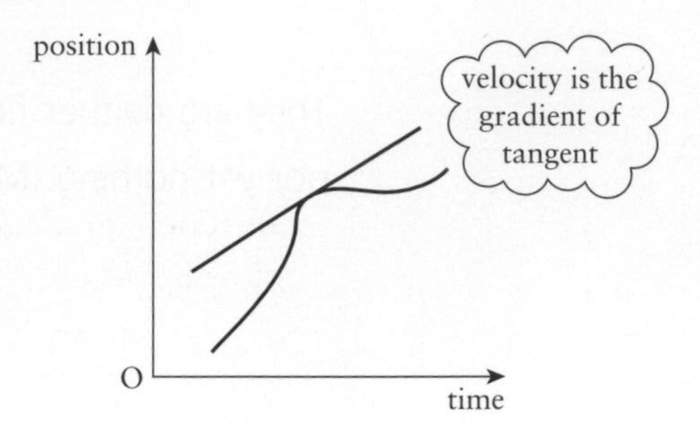

FIGURE 2.2

Similarly, the acceleration is the rate at which the velocity changes, so

$$a = \frac{dv}{dt} = \frac{d^2s}{dt^2} \quad ②$$

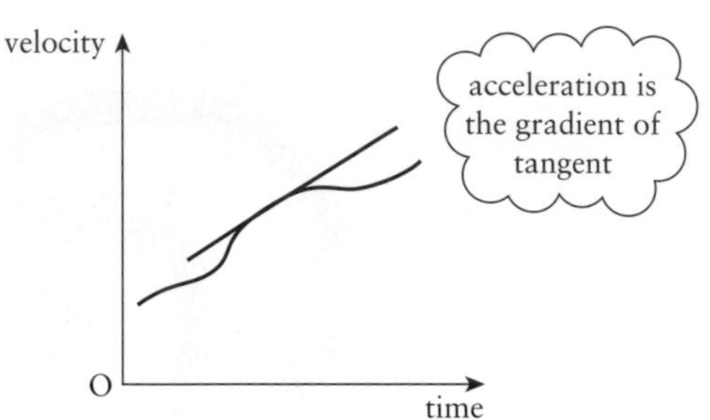

FIGURE 2.3

USING DIFFERENTIATION

When you are given the position of a moving object in terms of time, you can use equations ① and ② to solve problems even when the acceleration is not constant.

EXAMPLE 2.1

An object moves along a straight line so that its position at time t in seconds is given by

$$x = 2t^3 - 6t \text{ (in metres)} \quad (t \geqslant 0).$$

(a) Find expressions for the velocity and acceleration of the object at time t.

(b) Find the values of x, v and a when $t = 0$, 1, 2 and 3.

(c) Sketch the graphs of x, v and a against time.

(d) Describe the motion of the object.

Solution

(a) Position $\quad x = 2t^3 - 6t \quad ①$

Velocity $\quad v = \frac{dx}{dt} = 6t^2 - 6 \quad ②$

Acceleration $\quad a = \frac{dv}{dt} = 12t \quad ③$

You can now use these three equations to solve problems about the motion of the object.

(b)

When	$t =$	0	1	2	3
From ①	$x =$	0	−4	4	36
From ②	$v =$	−6	0	18	48
From ③	$a =$	0	12	24	36

(c) The graphs are drawn under each other so that you can see how they relate.

(d) The object starts at the origin and moves towards the negative direction, gradually slowing down.

At $t = 1$ it stops instantaneously and changes direction, returning to its initial position at about $t = 1.7$.

It then continues moving in the positive direction with increasing speed.

The acceleration is increasing at a constant rate. This cannot go on for much longer or the speed will become excessive.

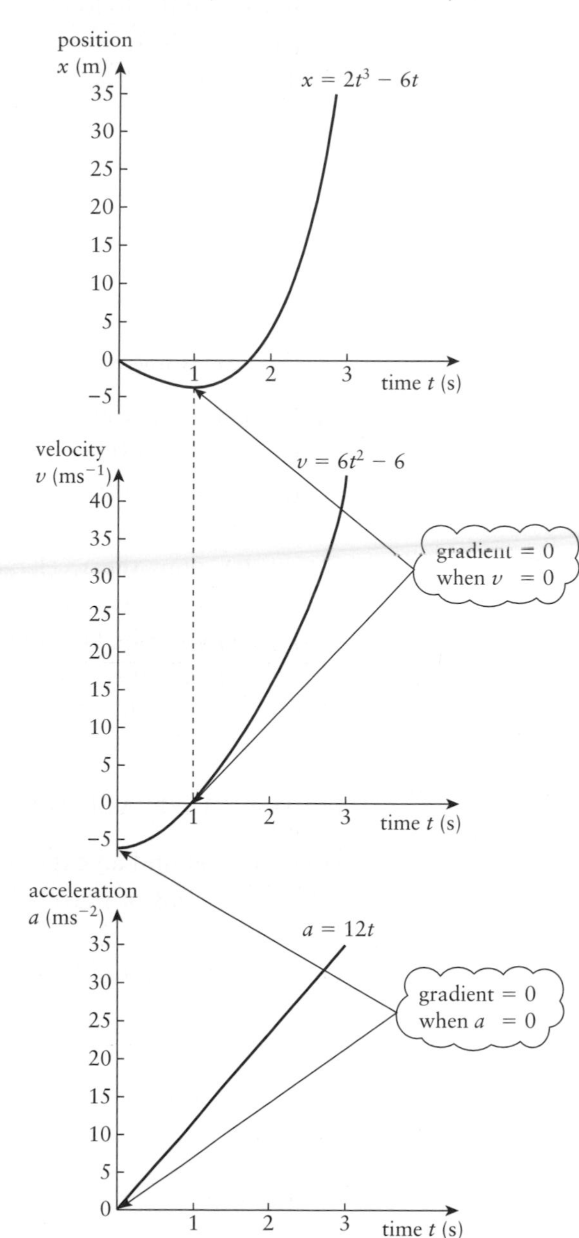

FIGURE 2.4

EXERCISE 2A

1 In each of the following cases

(a) $x = 10 + 2t - t^2$

(b) $x = -4t + t^2$

(c) $x = t^3 - 5t^2 + 4$

(i) find expressions for the velocity

(ii) use your equations to write down the initial position and velocity

(iii) find the time and position when the velocity is zero.

2 In each of the following cases

(a) $v = 4t + 3$

(b) $v = 6t^2 - 2t + 1$

(c) $v = 7t - 5$

(i) find expressions for the acceleration

(ii) use your equations to write down the initial velocity and acceleration.

3 The distance travelled by a cyclist is modelled by

$x = 4t + 0.5t^2$ in S.I. units.

Find expressions for the velocity and the acceleration of the cyclist at time t.

4 In each of the following cases

(a) $x = 15t - 5t^2$

(b) $x = 6t^3 - 18t^2 - 6t + 3$

(i) find expressions for the velocity and the acceleration

(ii) draw the acceleration–time graph and, below it, the velocity–time graph with the same scale for time and the origins in line

(iii) describe how the two graphs for each object relate to each other

(iv) describe how the velocity and acceleration change during the motion of each object.

FINDING DISPLACEMENT FROM VELOCITY

How can you find an expression for the position of an object when you know its velocity in terms of time?

One way of thinking about this is to remember that $v = \frac{\mathrm{d}s}{\mathrm{d}t}$, so you need to do the opposite of differentiation, that is integrate, to find s.

$$s = \int v \, \mathrm{d}t$$

The dt indicates that you must write v in terms of t before integrating

EXAMPLE 2.2

The velocity (in ms^{-1}) of a model train which is moving along straight rails is

$$v = 0.3t^2 - 0.5$$

Find its displacement from its initial position

(a) after time t
(b) after 3 seconds.

Solution

(a) The displacement at any time is $s = \int v\,dt$

$$= \int (0.3t^2 - 0.5)\,dt$$
$$= 0.1t^3 - 0.5t + c$$

To find the train's displacement from its initial position, put $s = 0$ when $t = 0$.

This gives $c = 0$ and so $s = 0.1t^3 - 0.5t$.

You can use this equation to find the displacement at any time before the motion changes.

(b) After 3 seconds, $t = 3$ and $s = 2.7 - 1.5$.

The train is 1.2 m from its initial position.

When using integration don't forget the constant. This is very important in mechanics problems and you are usually given some extra information to help you find the value of the constant.

THE AREA UNDER A VELOCITY–TIME GRAPH

In *Mechanics 1* you saw that the area under a velocity–time graph represents a displacement. Both the area under the graph and the displacement are found by integrating. To find a particular displacement you calculate the area under the velocity–time graph by integration using suitable limits.

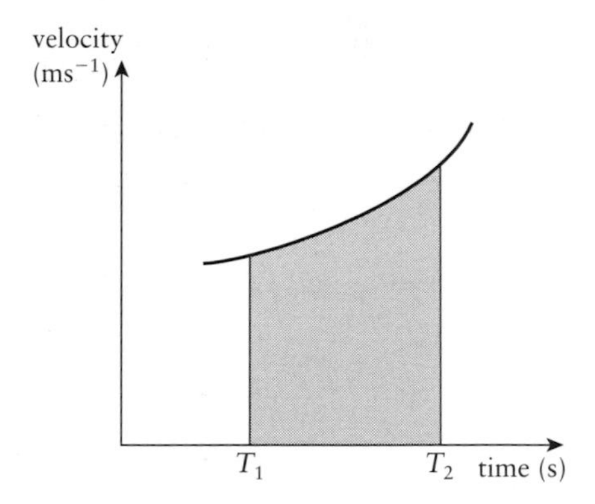

FIGURE 2.5

The distance travelled between the times T_1 and T_2 is shown by the shaded area on the graph.

$$s = \text{area} = \int_{T_1}^{T_2} v\,dt$$

EXAMPLE 2.3

A car moves between two sets of traffic lights, stopping at both. Its speed $v\,\text{ms}^{-1}$ at time t s is modelled by

$$v = \frac{1}{20}t(40 - t), \quad 0 \leqslant t \leqslant 40$$

Find the times at which the car is stationary and the distance between the two sets of traffic lights.

Solution The car is stationary when $v = 0$. Substituting this into the expression for the speed gives

$$0 = \frac{1}{20}t(40 - t)$$

$$t = 0 \text{ or } t = 40$$

These are the times when the car starts to move away from the first set of traffic lights and stops at the second set.

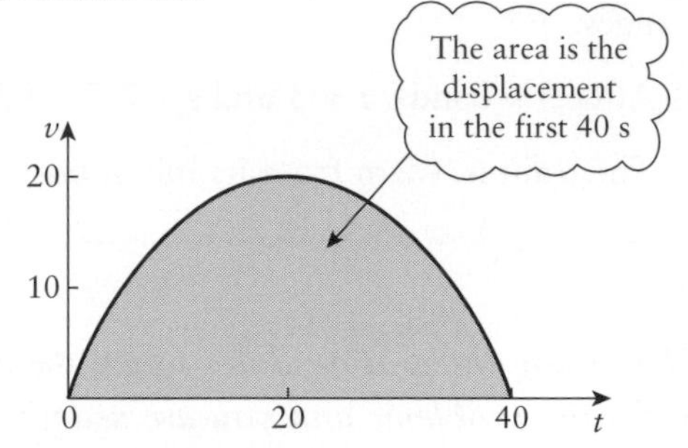

FIGURE 2.6

The distance between the two sets of lights is given by

$$\text{Distance} = \int_0^{40} \tfrac{1}{20}t(40 - t)\,\text{d}t$$

$$= \tfrac{1}{20}\int_0^{40} 40t - t^2\,\text{d}t$$

$$= \tfrac{1}{20}\left[20t^2 - \tfrac{t^3}{3}\right]_0^{40}$$

$$= 533.\dot{3}\text{ m}$$

FINDING VELOCITY FROM ACCELERATION

You can also find the velocity from the acceleration by using integration.

$$a = \frac{\text{d}v}{\text{d}t}$$

$$\Rightarrow \quad v = \int a\,\text{d}t$$

The next example shows how you can obtain equations for motion using integration.

EXAMPLE 2.4

The acceleration of a particle (in ms^{-2}) at time t seconds is given by

$$a = 6 - t$$

The particle is initially at the origin with velocity $-2\,ms^{-1}$. Find an expression for

(a) the velocity of the particle after t s

(b) the position of the particle after t s.

Hence find the velocity and position 6 s later.

Solution The information given may be summarised as follows:

at $t = 0$, $s = 0$ and $v = -2$;

at time t, $a = 6 - t$. ①

(a) $$\frac{dv}{dt} = a = 6 - t$$

Integrating gives

$$v = 6t - \tfrac{1}{2}t^2 + c$$

When $t = 0$, $v = -2$

so $-2 = 0 - 0 + c$

$c = -2$

at time t

$$v = 6t - \tfrac{1}{2}t^2 - 2 \quad ②$$

(b) $$\frac{ds}{dt} = v = 6t - \frac{1}{2}t^2 - 2$$

Integrating gives

$$s = 3t^2 - \tfrac{1}{6}t^3 - 2t + k$$

When $t = 0$, $s = 0$

so $0 = 0 - 0 - 0 + k$

$k = 0$

At time t

$$s = 3t^2 - \tfrac{1}{6}t^3 - 2t \quad ③$$

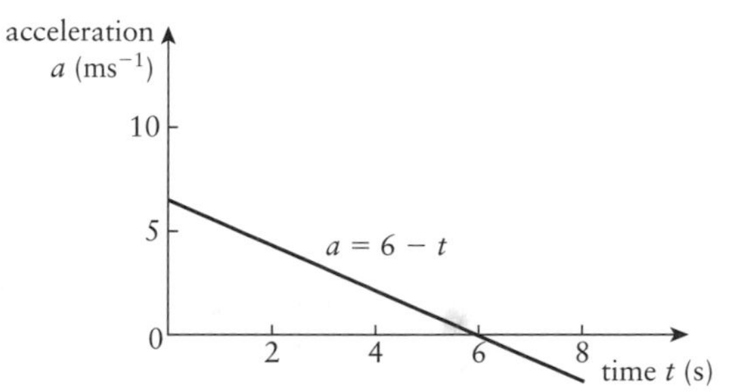

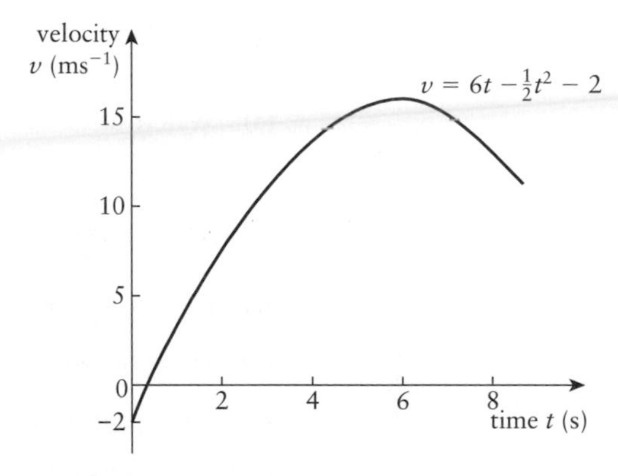

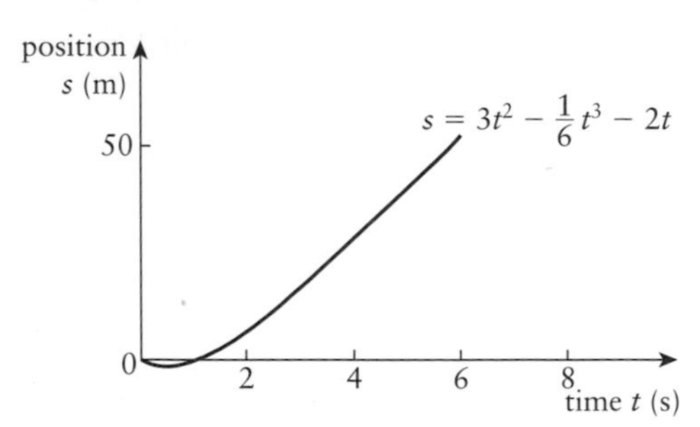

FIGURE 2.7

Notice that two different arbitrary constants (c and k) are necessary when you integrate twice. You could call them c_1 and c_2 if you wish.

The three numbered equations can now be used to give more information about the motion in a similar way to the *uvast* equations. (The *uvast* equations only apply when the acceleration is constant.)

When $t = 6$ $\quad v = 36 - 18 - 2 = 16$ from ②

When $t = 6$ $\quad s = 108 - 36 - 12 = 60$ from ③

The particle has a velocity of $+16\,\text{ms}^{-1}$ and is at $+60$ m after 6 s.

EXERCISE 2B

1 Find expressions for the position in each of these cases.

(a) $v = 4t + 3$; initial position 0.

(b) $v = 6t^3 - 2t^2 + 1$; when $t = 0$, $s = 1$.

(c) $v = 7t^2 - 5$; when $t = 0$, $s = 2$.

2 The speed of a ball rolling down a hill is modelled by $v = 1.7t$ (in ms^{-1}).

(a) Draw the speed–time graph of the ball.

(b) How far does the ball travel in 10 s?

3 Until it stops moving, the speed of a bullet t s after entering water is modelled by $v = 216 - t^3$ (in ms^{-1}).

(a) When does the bullet stop moving?

(b) How far has it travelled by this time?

4 During braking the speed of a car is modelled by $v = 40 - 2t^2$ (in ms^{-1}) until it stops moving.

(a) How long does the car take to stop?

(b) How far does it move before it stops?

5 In each case below, the object moves along a straight line with acceleration a in ms^{-2}. Find an expression for the velocity v (ms^{-1}) and position x (m) of each object at time t s.

(a) $a = 10 + 3t - t^2$; the object is initially at the origin and at rest.

(b) $a = 4t - 2t^2$; at $t = 0$, $x = 1$ and $v = 2$.

(c) $a = 10 - 6t$; at $t = 1$, $x = 0$ and $v = -5$.

6 A boy throws a ball up in the air from a height of 1.5 m and catches it at the same height. Its height in metres at time t seconds is

$$y = 1.5 + 15t - 5t^2.$$

(a) What is the vertical velocity $v\,\text{ms}^{-1}$ of the ball at time t?

(b) Find the position, velocity and speed of the ball at $t = 1$ and $t = 2$.

(c) Sketch the position–time, velocity–time and speed–time graphs for $0 \leqslant t \leqslant 3$.

(d) When does the boy catch the ball?

(e) Explain why the distance travelled by the ball is not equal to $\int_0^3 v\,dt$ and state what information this expression does give.

7 An object moves along a straight line so that its position in metres at time t seconds is given by

$$x = t^3 - 3t^2 - t + 3 \quad (t \geqslant 0).$$

(a) Find the position, velocity and speed of the object at $t = 2$.

(b) Find the smallest time when

(i) the position is zero

(ii) the velocity is zero.

(c) Sketch position–time, velocity–time and speed–time graphs for $0 \leqslant t \leqslant 3$.

(d) Describe the motion of the object.

8 Two objects move along the same straight line. The velocities of the objects (in ms^{-1}) are given by $v_1 = 16t - 6t^2$ and $v_2 = 2t - 10\,(t \geqslant 0)$.

Initially the objects are 32 m apart. At what time do they collide?

9 An object moves along a straight line so that its acceleration in metres per second squared is given by $a = 4 - 2t$. It starts its motion at the origin with speed $4\,\text{ms}^{-1}$ in the direction of x increasing.

(a) Find as functions of t the velocity and position of the object.

(b) Sketch the position–time, velocity–time and acceleration–time graphs for $0 \leqslant t \leqslant 2$.

(c) Describe the motion of the object.

10 Nick watches a golfer putting her ball 24 m from the edge of the green and into the hole and he decides to model the motion of the ball. Assuming that the ball is a particle travelling along a straight line he models its distance, s metres, from the golfer at time t seconds by

$$s = -\tfrac{3}{2}t^2 + 12t \quad 0 \leqslant t \leqslant 4.$$

(a) Find the value of s when $t = 0$, 1, 2, 3 and 4.

(b) Explain the restriction $0 \leqslant t \leqslant 4$.

(c) Find the velocity of the ball at time t seconds.

(d) With what speed does the ball enter the hole?

(e) Find the acceleration of the ball at time t seconds.

11 Andrew and Elizabeth are having a race over 100 m. Their accelerations (in ms^{-2}) are as follows:

Andrew		Elizabeth	
$a = 4 - 0.8t$	$0 \leqslant t \leqslant 5$	$a = 4$	$0 \leqslant t \leqslant 2.4$
$a = 0$	$t > 5$	$a = 0$	$t > 2.4$

(a) Find the greatest speed of each runner.
(b) Sketch the speed–time graph for each runner.
(c) Find the distance Elizabeth runs while reaching her greatest speed.
(d) How long does Elizabeth take to complete the race?
(e) Who wins the race, by what time margin and by what distance?

On another day they race over 120 m, both running in exactly the same manner.

(f) What is the result now?

12 Christine is a parachutist. On one of her descents her vertical speed, $v\,\text{ms}^{-1}$, t s after leaving an aircraft is modelled by

$$\begin{aligned} v &= 8.5t && 0 \leqslant t \leqslant 10 \\ v &= 5 + 0.8(t - 20)^2 && 10 < t \leqslant 20 \\ v &= 5 && 20 < t \leqslant 90 \\ v &= 0 && t > 90 \end{aligned}$$

(a) Sketch the speed–time graph for Christine's descent and explain the shape of each section.
(b) How high is the aircraft when Christine jumps out?
(c) Write down expressions for the acceleration during the various phases of Christine's descent. What is the greatest magnitude of her acceleration?

13 A train starts from rest at a station. Its acceleration is shown on the acceleration–time graph below.

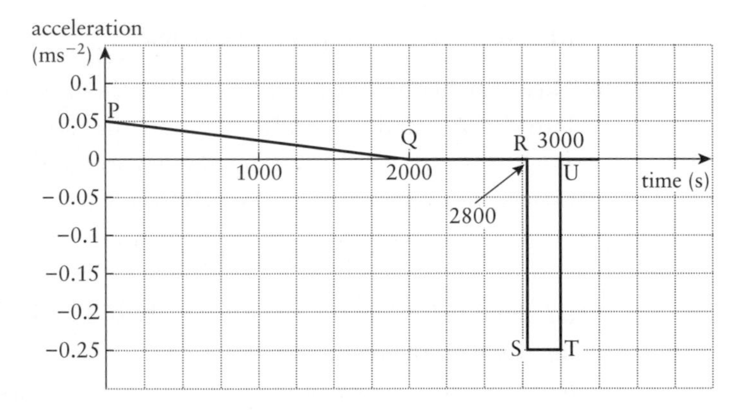

(a) Describe what is happening during the phases of the train's journey represented by the lines PQ, QR and ST.
(b) The equation of the line PQ is of the form $a = mt + c$. Find the values of the constants m and c.
(c) Find the maximum speed of the train.
(d) What is the speed of the train when $t = 3000$?
(e) How far does the train travel during the first 3000 s?

Motion in two dimensions

In your work on projectile motion you have met the idea that the position of an object can be represented by a vector

$$\mathbf{r} = x\mathbf{i} + y\mathbf{j}$$

When a ball is thrown into the air, for example, its position might be given by

$$\mathbf{r} = 5t\mathbf{i} + (12t - 5t^2)\mathbf{j}$$

so in this case $x = 5t$ and $y = 12t - 5t^2$. You can plot the path of the ball by finding the values of $\mathbf{r}$ and hence x and y for several values of t.

t	0	0.5	1	1.5	2	2.4
$\mathbf{r}$	0	$2.5\mathbf{i} + 4.75\mathbf{j}$	$5\mathbf{i} + 7\mathbf{j}$	$7.5\mathbf{i} + 6.75\mathbf{j}$	$10\mathbf{i} + 4\mathbf{j}$	$12\mathbf{i}$

Figure 2.8 shows the path of the ball and also its position $\mathbf{r}$ when $t = 2$.

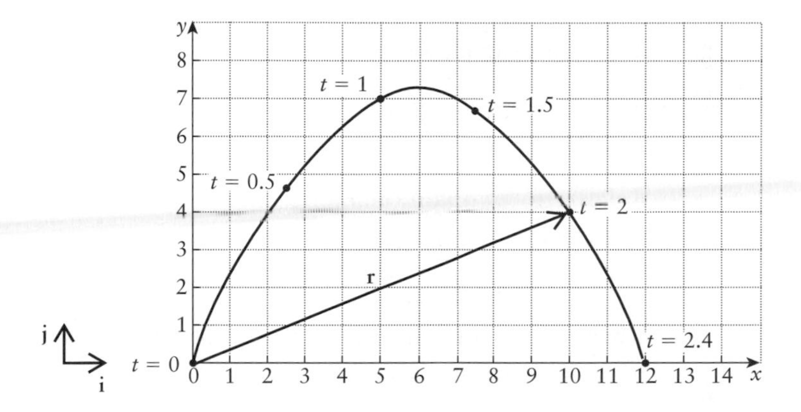

Figure 2.8

Finding the velocity and acceleration

The equation for $\mathbf{r}$ can be differentiated to give the velocity and acceleration.

When $\quad \mathbf{r} = 5t\mathbf{i} + (12t - 5t^2)\mathbf{j}$

$$\mathbf{v} = 5\mathbf{i} + (12 - 10t)\mathbf{j}$$

$$\mathbf{a} = -10\mathbf{j}$$

Note The direction of the acceleration is not at all obvious when you look at the diagram of the path of the ball.

NEWTON'S NOTATION

When you write derivatives in vectors, the notation becomes very cumbersome so many people use Newton's notation when differentiating with respect to time. In this notation a dot is placed over the variable for each differentiation. For example $\dot{x} = \frac{dx}{dt}$ and $\ddot{x} = \frac{d^2x}{dt^2}$.

$$\mathbf{v} = \dot{\mathbf{r}} = \dot{x}\mathbf{i} + \dot{y}\mathbf{j}$$

Speed
$|\mathbf{v}| = \sqrt{\dot{x}^2 + \dot{y}^2}$
Direction of motion
$\tan\alpha = \frac{\dot{y}}{\dot{x}}$

$$\mathbf{a} = \dot{\mathbf{v}} = \ddot{\mathbf{r}}$$

$$\mathbf{a} = \ddot{x}\mathbf{i} + \ddot{y}\mathbf{j}$$

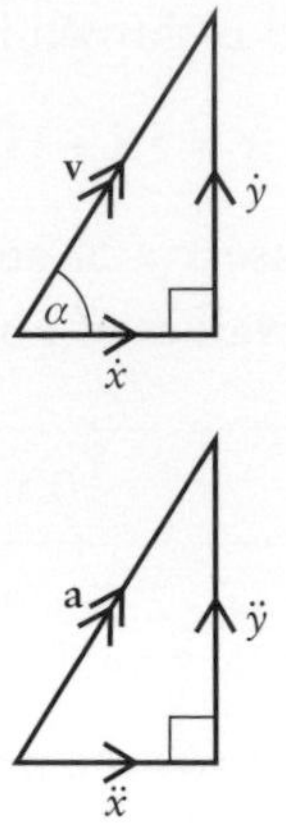

FIGURE 2.9

The next example shows how you can use these results.

EXAMPLE 2.5

Relative to an origin on a long, straight beach, the position of a speedboat is modelled by the vector

$$\mathbf{r} = (2t + 2)\mathbf{i} + (12 - t^2)\mathbf{j}$$

where **i** and **j** are unit vectors perpendicular and parallel to the beach. Distances are in metres and the time t is in seconds.

(a) Calculate the distance of the boat from the origin, O, when the boat is 6 m from the beach.

(b) Sketch the path of the speedboat for $0 \leqslant t \leqslant 3$.

(c) Find expressions for the velocity and acceleration of the speedboat at time t. Is the boat ever at rest? Explain your answer.

(iv) For $t = 3$, calculate the speed of the boat and the angle its direction of motion makes to the line of the beach.

(e) Suggest why this model for the motion of the speedboat is unrealistic for large t.

[MEI]

Solution

(a) $\mathbf{r} = (2t + 2)\mathbf{i} + (12 - t^2)\mathbf{j}$ so the boat is 6 m from the beach when $x = 2t + 2 = 6$

Then $t = 2$ and $y = 12 - t^2 = 8$

The distance from O is $\sqrt{6^2 + 8^2} = 10$ m.

(b) The table shows the position at different times and the path of the boat is shown on the graph in figure 2.10.

t	0	1	2	3
$\mathbf{r}$	$2\mathbf{i} + 12\mathbf{j}$	$4\mathbf{i} + 11\mathbf{j}$	$6\mathbf{i} + 8\mathbf{j}$	$8\mathbf{i} + 3\mathbf{j}$

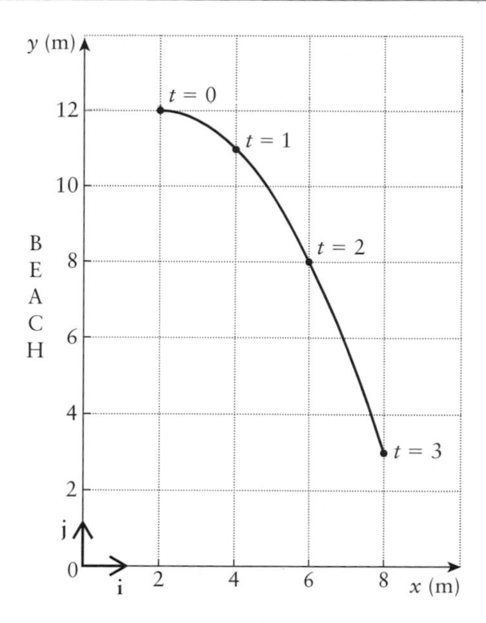

FIGURE 2.10

(c)
$$\mathbf{r} = (2t + 2)\mathbf{i} + (12 - t^2)\mathbf{j}$$
$$\Rightarrow \quad \mathbf{v} = \dot{\mathbf{r}} = 2\mathbf{i} - 2t\mathbf{j}$$
$$\text{and} \quad \mathbf{a} = \dot{\mathbf{v}} = -2\mathbf{j}$$

The boat is at rest if both components of velocity ($\dot{x}$ and $\dot{y}$) are zero at the same time. But $\dot{x}$ is always 2, so the velocity can never be zero.

(d) When $t = 3$ $\quad \mathbf{v} = 2\mathbf{i} - 6\mathbf{j}$

The angle $\mathbf{v}$ makes with the beach is α as shown where

$$\tan\alpha = \frac{2}{6}$$
$$\alpha = 18.4°$$

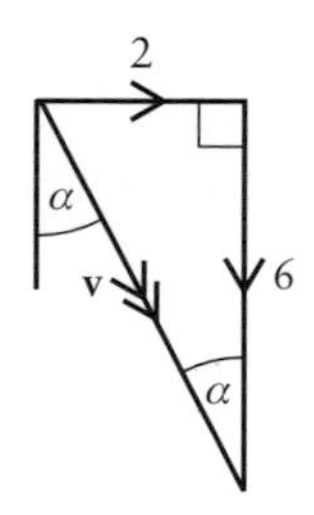

FIGURE 2.11

(e) According to this model, the speed after time t is

$$|\mathbf{v}| = |2\mathbf{i} - 2t\mathbf{j}| = \sqrt{2^2 + (-2t)^2} = \sqrt{4 + 4t^2}$$

As t increases, the speed increases at an increasing rate so there must come a time when the boat is incapable of going at the predicted speed and the model cannot then apply.

Notice that the direction of motion is found using the velocity and not the position.

USING INTEGRATION

When you are given the velocity or acceleration and wish to work backwards to the displacement, you need to integrate. The next two examples show how you can do this with vectors.

EXAMPLE 2.6

An aircraft is dropping a crate of supplies on to level ground. Relative to an observer on the ground, the crate is released at the point with position vector $650\mathbf{i} + 576\mathbf{j}$ m and with initial velocity $-100\mathbf{i}\,\text{ms}^{-1}$, where the directions are horizontal and vertical. Its acceleration is modelled by

$$\mathbf{a} = (-t + 12)\mathbf{i} + \left(\tfrac{1}{2}t - 10\right)\mathbf{j} \text{ for } t \leqslant 12\,\text{s}$$

(a) Find an expression for the velocity vector of the crate at time t.
(b) Find an expression for the position vector of the crate at time t.
(c) Verify that the crate hits the ground 12 s after its release and find how far from the observer this happens.

Solution **(a)**

$$\mathbf{a} = \frac{d\mathbf{v}}{dt} = (-t + 12)\mathbf{i} + \left(\tfrac{1}{2}t - 10\right)\mathbf{j} \quad ①$$

You can treat horizontal and vertical motion separately if you wish

Integrating gives $\mathbf{v} = \left(-\tfrac{1}{2}t^2 + 12t + c_1\right)\mathbf{i} + \left(\tfrac{1}{4}t^2 - 10t + c_2\right)\mathbf{j}$

At $t = 0$ $\quad \mathbf{v} = -100\mathbf{i} \quad \Rightarrow \quad 0 + 0 + c_1 = -100$

$\quad 0 - 0 + c_2 = 0$

So velocity $\quad \mathbf{v} = \left(-\tfrac{1}{2}t^2 + 12t - 100\right)\mathbf{i} + \left(\tfrac{1}{4}t^2 - 10t\right)\mathbf{j} \quad ②$

(b)

$$\mathbf{v} = \frac{d\mathbf{r}}{dt}$$

Integrating again gives $\mathbf{r} = \left(-\tfrac{1}{6}t^3 + 6t^2 - 100t + k_1\right)\mathbf{i} + \left(\tfrac{1}{12}t^3 - 5t^2 + k_2\right)\mathbf{j}$

At $t = 0$ $\quad \mathbf{r} = 650\mathbf{i} + 576\mathbf{j} \quad \Rightarrow \quad k_1 = 650$

$\quad k_2 = 576$

Position vector $\quad \mathbf{r} = \left(-\tfrac{1}{6}t^3 + 6t^2 - 100t + 650\right)\mathbf{i} + \left(\tfrac{1}{12}t^3 - 5t^2 + 576\right)\mathbf{j} \quad ③$

(c) At t = 12

$$\mathbf{r} = \left(-\tfrac{1}{6} \times 12^3 + 6 \times 12^2 - 100 \times 12 + 650\right)\mathbf{i} + \left(\tfrac{1}{12} \times 12^3 - 5 \times 12^2 + 576\right)\mathbf{j}$$

$$\mathbf{r} = 26\mathbf{i}$$

Since $y = 0$, the crate hits the ground after 12 s and it is then $x = 26$ m in front of the observer.

Note When you integrate a vector in two dimensions you need a constant of integration for each direction, for example c_1 and c_2 as above.

It is also a good idea to number your equations for **a**, **v** and **r** so that you can find them easily if you want to use them later.

FORCE AS A FUNCTION OF TIME

When the force acting on an object is given as a function of t you can use Newton's second law to find out about its motion. You can now write this as $\mathbf{F} = m\mathbf{a}$ because force and acceleration are both vectors.

EXAMPLE 2.7

A force of $(12\mathbf{i} + 3t\mathbf{j})$ N, where t is the time in seconds, acts on a particle of mass 6 kg. The directions of **i** and **j** correspond to east and north respectively.

(a) Show that the acceleration is $(2\mathbf{i} + 0.5t\mathbf{j})\,\text{ms}^{-2}$ at time t.
(b) Find the acceleration and magnitude of the acceleration when $t = 12$.
(c) At what time is the acceleration directed north-east (i.e. a bearing of 045°)?
(d) If the particle starts with a velocity of $(2\mathbf{i} - 3\mathbf{j})\,\text{ms}^{-1}$ when $t = 0$, what will its velocity be when $t = 3$?
(e) When $t = 3$ a second **constant** force begins to act. Given that the acceleration of the particle at that time due to both forces is $4\,\text{ms}^{-2}$ due south, find the second force.

[MEI]

Solution **(a)** By Newton's second law the force = mass × acceleration

$$(12\mathbf{i} + 3t\mathbf{j}) = 6\mathbf{a}$$
$$\mathbf{a} = \tfrac{1}{6}(12\mathbf{i} + 3t\mathbf{j})$$
$$\mathbf{a} = 2\mathbf{i} + 0.5t\mathbf{j} \quad ①$$

(b) When $t = 12$ $\qquad \mathbf{a} = 2\mathbf{i} + 6\mathbf{j}$

magnitude of **a** $\qquad |\mathbf{a}| = \sqrt{2^2 + 6^2}$

The acceleration is $2\mathbf{i} + 6\mathbf{j}\,\text{ms}^{-2}$ with magnitude $6.32\,\text{ms}^{-2}$.

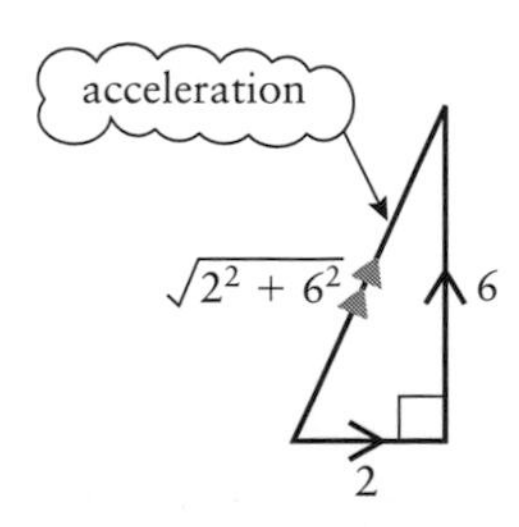

FIGURE 2.12

(c) The acceleration is north-east when its northerly component is equal to its easterly component. From ①, this is when $2 = 0.5t$ i.e. when $t = 4$.

(d) The velocity at time t is $\int \mathbf{a}\,dt = \int (2\mathbf{i} + 0.5t\mathbf{j})\,dt$

$$\Rightarrow \quad \mathbf{v} = 2t\mathbf{i} + 0.25t^2\mathbf{j} + \mathbf{c}$$

The constant **c** is a vector such as $c_1\mathbf{i} + c_2\mathbf{j}$

When $t = 0$, $\mathbf{v} = 2\mathbf{i} - 3\mathbf{j}$

so $\quad 2\mathbf{i} - 3\mathbf{j} = 0\mathbf{i} + 0\mathbf{j} + \mathbf{c} \Rightarrow \mathbf{c} = 2\mathbf{i} - 3\mathbf{j}$

$$\mathbf{v} = 2t\mathbf{i} + 0.25t^2\mathbf{j} + 2\mathbf{i} - 3\mathbf{j}$$

$$\mathbf{v} = (2t + 2)\mathbf{i} + (0.25t^2 - 3)\mathbf{j} \qquad ②$$

When $t = 3$ $\quad \mathbf{v} = 8\mathbf{i} - 0.75\mathbf{j}$

(e) Let the second force be **F** so the total force when $t = 3$ is $(12\mathbf{i} + 3 \times 3\mathbf{j}) + \mathbf{F}$.

The acceleration is $-4\mathbf{j}$, so by Newton II

$$(12\mathbf{i} + 9\mathbf{j}) + \mathbf{F} = 6 \times -4\mathbf{j}$$

$$\mathbf{F} = -24\mathbf{j} - 12\mathbf{i} - 9\mathbf{j}$$

$$\mathbf{F} = -12\mathbf{i} - 33\mathbf{j}$$

The second force is $-12\mathbf{i} - 33\mathbf{j}$ N.

Note You can use the same methods in three dimensions just by including a third direction **k**, for example

$\mathbf{r} = x\mathbf{i} + y\mathbf{j} + z\mathbf{k}$

$\mathbf{v} = \dot{x}\mathbf{i} + \dot{y}\mathbf{j} + \dot{z}\mathbf{k}$

$\mathbf{a} = \ddot{x}\mathbf{i} + \ddot{y}\mathbf{j} + \ddot{z}\mathbf{k}$.

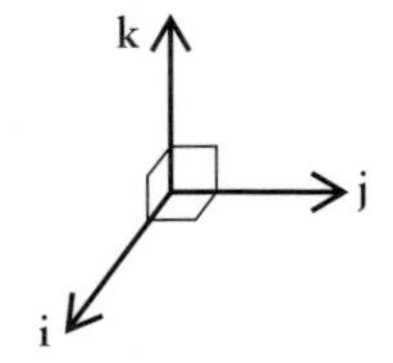

FIGURE 2.13

Historical Note Newton's work on motion required more mathematical tools than were generally used at the time. He had to invent his own ways of thinking about continuous change and in about 1666 he produced a theory of 'fluxions' in which he imagined a quantity 'flowing' from one magnitude to another. This was the beginning of calculus. He did not publish his methods, however, and when Leibniz published his version in 1684 there was an enormous amount of controversy amongst their supporters about who was first to discover calculus. The sharing of ideas between mathematicians in Britain and the rest of Europe was hindered for a century. The contributions of both men are remembered today by their notation. Leibniz's $\frac{dx}{dt}$ is common and Newton's $\dot{x}$ is widely used in mechanics.

EXERCISE 2C

1 The first part of a race track is a bend. As the leading car travels round the bend its position, in metres, is modelled by

$$\mathbf{r} = 2t^2\mathbf{i} + 8t\mathbf{j}$$

where t is in seconds.

(a) Find an expression for the velocity of the car.

(b) Find the position of the car when $t = 0, 1, 2, 3$ and 4.
Use this information to sketch the path of the car.

(c) Find the velocity of the car when $t = 0, 1, 2, 3$ and 4.
Add vectors to your sketch to represent these velocities.

(d) Find the speed of the car as it leaves the bend at $t = 5$.

2 As a boy slides down a slide his position vector in metres at time t is

$$\mathbf{r} = (16 - 4t)\mathbf{i} + (20 - 5t)\mathbf{j}$$

Find his velocity and acceleration.

3 Calculate the magnitude and direction of the acceleration of a particle that moves so that its position vector in metres is given by

$$\mathbf{r} = (8t - 2t^2)\mathbf{i} + (6 + 4t - t^2)\mathbf{j}$$

where t is the time in seconds.

4 A rocket moves with a velocity (in ms^{-1}) modelled by

$$\mathbf{v} = \tfrac{1}{10}t\mathbf{i} + \tfrac{1}{10}t^2\mathbf{j}$$

where $\mathbf{i}$ and $\mathbf{j}$ are horizontal and vertical unit vectors respectively and t is in seconds. Find

(a) an expression for its position vector relative to its starting position at time t

(b) the displacement of the rocket after 10 s of its flight.

5 A particle is initially at rest at the origin. It experiences an acceleration given by

$$\mathbf{a} = 4t\mathbf{i} + (6 - 2t)\mathbf{j}.$$

Find expressions for the velocity and position of the particle at time t.

6 While a hockey ball is being hit it experiences an acceleration (in ms^{-2}) modelled by

$$\mathbf{a} = 1000[6t(t - 0.2)\mathbf{i} + t(t - 0.2)\mathbf{j}] \text{ for } 0 \leqslant t \leqslant 0.2 \text{ in seconds}$$

If the ball is initially at rest, find its speed when it loses contact with the stick after 0.2 s.

7 A speedboat is initially moving at $5\,\text{ms}^{-1}$ on a bearing of $135°$.

(a) Express the initial velocity as a vector in terms of **i** and **j**, which are unit vectors east and north respectively.

The boat then begins to accelerate with an acceleration modelled by

$$\mathbf{a} = 0.1t\mathbf{i} + 0.3t\mathbf{j} \quad \text{in ms}^{-2}$$

(b) Find the velocity of the boat 10 s after it begins to accelerate and its displacement over the 10 s period.

8 A girl throws a ball and, t seconds after she releases it, its position in metres relative to the point where she is standing is modelled by

$$\mathbf{r} = 15t\mathbf{i} + (2 + 16t - 5t^2)\mathbf{j}$$

where the directions are horizontal and vertical.

(a) Find expressions for the velocity and acceleration of the ball at time t.

(b) The vertical component of the velocity is zero when the ball is at its highest point. Find the time taken for the ball to reach this point.

(c) When the ball hits the ground the vertical component of its position vector is zero. What is the speed of the ball when it hits the ground?

9 The position (in metres) of a tennis ball t seconds after leaving a racquet is modelled by

$$\mathbf{r} = 20t\mathbf{i} + (2 + t - 5t^2)\mathbf{j}$$

where **i** and **j** are horizontal and vertical unit vectors.

(a) Find the position of the tennis ball when $t = 0, 0.2, 0.4, 0.6$ and 0.8. Use these to sketch the path of the ball.

(b) Find an expression for the velocity of the tennis ball. Use this to find the velocity of the ball when $t = 0.2$.

(c) Find the acceleration of the ball.

10 An owl is initially perched on a tree. It then goes for a short flight which ends when it dives on to a mouse on the ground. The position vector (in metres) of the owl t seconds into its flight is modelled by

$$\mathbf{r} = t^2(6 - t)\mathbf{i} + (12.5 + 4.5t^2 - t^3)\mathbf{j}$$

where the foot of the tree is taken to be the origin and the unit vectors **i** and **j** are horizontal and vertical.

(a) Draw a graph showing the bird's flight.

(b) For how long (in s) is the owl in flight?

(c) Find the speed of the owl when it catches the mouse and the angle that its flight makes with the horizontal at that instant.

(d) Show that the owl's acceleration is never zero during the flight.

11 Ship A is 5 km due west of ship B and is travelling on a course 035° at a constant but unknown speed $v\,\text{km h}^{-1}$. Ship B is travelling at a constant $10\,\text{km h}^{-1}$ on a course 300°.

(a) Write the velocity of each ship in terms of unit vectors **i** and **j** with directions east and north.

(b) Find the position vector of each ship at time t hours, relative to the starting position of ship A.

The ships are on a collision course.

(c) Find the speed of ship A.

(d) How much time elapses before the collision occurs?

12 A particle of mass 0.5 kg is acted on by a force, in newtons,

$$\mathbf{F} = t^2\mathbf{i} + 2t\mathbf{j}.$$

The particle is initially at rest at the origin and t is measured in seconds.

(a) Find the acceleration of the particle at time t.

(b) Find the velocity of the particle at time t.

(c) Find the position vector of the particle at time t.

(d) Give all the information you can about the particle at time $t = 2$.

13 The position vector of a motorcycle of mass 150 kg on a track is modelled by

$$\mathbf{r} = 4t^2\mathbf{i} + \tfrac{1}{8}t(8-t)^2\mathbf{j} \qquad 0 \leqslant t \leqslant 8$$
$$\mathbf{r} = (64t - 256)\mathbf{i} \qquad 8 < t \leqslant 20$$

where t is the time in seconds after the start of a race.

The vectors **i** and **j** are in directions along and perpendicular to the direction of the track as shown in the diagram. The origin is in the middle of the track.

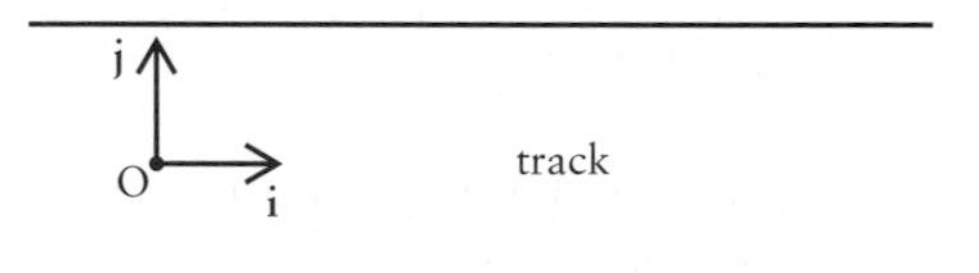

(a) Draw a sketch to show the motorcycle's path over the first 10 s. The track is 20 m wide. Does the motorcycle leave it?

(b) Find, in vector form, expressions for the velocity and acceleration of the motorcycle at time t for $0 \leqslant t \leqslant 20$.

(c) Find in vector form an expression for the resultant horizontal force acting on the motorcycle during the first 8 s, in terms of t.

(d) Why would you expect the driving force from the motorcycle's engine to be substantially greater than the component in the **i** direction of your answer to part **(c)**?

EXERCISE 2D **Examination-style questions**

1 A particle P is moving along the x axis. At time $t = 0$ it is at the point where $x = 3$ and moving with speed $4\,\text{ms}^{-1}$ in the direction of x increasing. At time t s the acceleration of P is $2t - 5\,\text{ms}^{-2}$ in the direction of x increasing.

(a) Find an expression for the velocity of P at time t.

(b) Find the position of P when it first comes to instantaneous rest.

2 A particle P moves along the x axis. Its position, x m, at time t s is given by

$$x = \tfrac{1}{12}t^3 - \tfrac{1}{2}t^2 - 4 \quad 0 \leqslant t \leqslant 6.$$

Find

(a) the values of t at which P is instantaneously at rest

(b) the distance between the points at which P is instantaneously at rest

(c) the maximum acceleration during the period $0 \leqslant t \leqslant 6$.

3 A particle P moves along the x axis. At time $t = 0$ it is at the origin O, and at time t it has velocity $v\,\text{ms}^{-1}$ given by

$$v = 3t^2 - 4t + 1.$$

Find

(a) the positions of P when its velocity is zero

(b) the time at which its acceleration is zero

(c) the total distance travelled by P in the first 2 s of its motion.

4 A lift is moving up a vertical shaft. It starts from rest at the bottom of the shaft, and its acceleration at time t s for the first 4 s of the motion is $a\,\text{ms}^{-2}$, where

$$a = 3t - 0.75t^2.$$

Find

(a) expressions in terms of t for the velocity $v\,\text{ms}^{-1}$ and the position x m at time t

(b) the maximum acceleration during the period $0 \leqslant t \leqslant 4$

(c) the maximum speed during the period $0 \leqslant t \leqslant 4$

(d) the total distance moved during the period $0 \leqslant t \leqslant 4$.

5 An athlete is running a 100-m race. The athlete starts from rest and the race is complete when the athlete has covered a distance of 100 m. In a first model, the athlete is assumed to have a constant acceleration of $2\,\text{ms}^{-2}$ for the first 6 s of the race and to move at constant velocity after that.

(a) Sketch a velocity–time graph and find the time taken for the athlete to finish the race using this model.

In a second model, the athlete is assumed to have acceleration $a\,\text{ms}^{-2}$ at time t s for the entire duration of the race, where $a = 3.5 - 0.5t$.

(b) Find expressions for the velocity and position at time t using the second model.

(c) Find the position of the athlete at time $t = 11$ using the second model and comment on your result.

(d) Find the times between which the athlete has a velocity greater than $12\,\text{ms}^{-1}$ using the second model.

6 A particle P is initially at the origin O and moves so that its position vector $\mathbf{r}$ m at time t s is given by

$$\mathbf{r} = (t^3 - 4t)\mathbf{i} + (1.5t^2 + t)\mathbf{j} \quad t \geqslant 0.$$

(a) Find the velocity of P at time t s.

(b) Hence find the time at which P is moving in a direction parallel to the vector $\mathbf{i} + 2\mathbf{j}$.

7 A particle P moves so that its velocity $\mathbf{v}\,\mathrm{m\,s^{-1}}$ at time t s is given by

$$\mathbf{v} = (2t - 6)\mathbf{i} + (3t - t^2)\mathbf{j}.$$

(a) Find an expression for its acceleration at time t.

At time $t = 0$, P is at the point with position vector $2\mathbf{i} - \mathbf{j}$ m.

(b) Find, in degrees to 1 decimal place, the angle between OP and the x axis at the moment when P is stationary.

8 A particle P is moving with acceleration $\mathbf{a}\,\mathrm{m\,s^{-2}}$ at time t s given by

$$\mathbf{a} = -2\mathbf{i} + (t - 1)\mathbf{j}, \quad t \geqslant 0.$$

At time $t = 0$, it is at the origin O, moving with velocity $4\mathbf{i} + \mathbf{j}\,\mathrm{m\,s^{-1}}$.

(a) Find an expression in terms of t for its velocity at a later time t.

(b) Find an expression in terms of t for its position at a later time t.

(c) At what time is P moving parallel to the vector $\mathbf{j}$?

9 A particle of mass 3 kg moves with an acceleration of $\mathbf{a}\,\mathrm{m\,s^{-2}}$ at time t s, given by

$$\mathbf{a} = (t^{\frac{1}{2}} + 1)\mathbf{i} + (2t - 3)\mathbf{j}$$

where $\mathbf{i}$ and $\mathbf{j}$ are unit vectors east and north respectively.

(a) Find the magnitude and direction of the resultant force acting on P when $t = 9$.

When $t = 0$, P has velocity $2\mathbf{j}\,\mathrm{m\,s^{-1}}$.

(b) Find an expression for the velocity of P at time t.

10 At time t s, a particle P of mass 0.5 kg is acted on a by a force $\mathbf{F}$ N, where

$$\mathbf{F} = (2 - t)\mathbf{i} + 3t^2\mathbf{j}$$

and $\mathbf{i}$ and $\mathbf{j}$ are unit vectors east and north respectively. At $t = 0$, P is at the origin and moving with velocity $-3\mathbf{i} - 2\mathbf{j}\,\mathrm{m\,s^{-1}}$.

(a) Find an expression in terms of t for the velocity of P at time t.

(b) Is the particle ever stationary? Give a reason for your answer.

(c) Find an expression in terms of t for the position of P at time t.

(d) At what positive time is the acceleration parallel to the vector $2\mathbf{i} + \mathbf{j}$?

11 A particle P moves on the x-axis. At time t seconds the velocity of P is $v\,\mathrm{m\,s^{-1}}$ in the direction of x increasing, where $v = 6t - 2t^2$. When $t = 0$, P is at the origin O. Find the distance of P from O when P comes to instantaneous rest after leaving O.

[Edexcel]

12 The velocity $\mathbf{v}\,\mathrm{m\,s^{-1}}$ of a particle P at time t seconds is given by

$$\mathbf{v} = (3t - 2)\mathbf{i} - 5t\mathbf{j}.$$

(a) Show that the acceleration of P is constant.

At $t = 0$, the position vector of P relative to a fixed origin O is $3\mathbf{i}$ m.

(b) Find the distance of P from O when $t = 2$.

[Edexcel]

13 A particle P moves on the x-axis. The acceleration of P at time t seconds is $(4t - 8)\,\mathrm{m\,s^{-2}}$, measured in the direction of x increasing. The velocity of P at time t seconds is $v\,\mathrm{m\,s^{-1}}$. Given that $v = 6$ when $t = 0$, find

(a) v in terms of t,

(b) the distance between the two points where P is instantaneously at rest.

[Edexcel]

14 At time t seconds, a particle P has position vector $\mathbf{r}$ metres relative to a fixed origin O, where

$$\mathbf{r} = (t^2 + 2t)\mathbf{i} + (t - 2t^2)\mathbf{j}.$$

Show the acceleration of P is constant and find its magnitude.

[Edexcel]

KEY POINTS

Relationships between the variables describing motion

	Position	→ Velocity	→ Acceleration
		differentiate →	
In one dimension	s	$v = \dfrac{ds}{dt}$	$a = \dfrac{dv}{dt} = \dfrac{d^2s}{dt^2}$
In two dimensions	$\mathbf{r} = x\mathbf{i} + y\mathbf{j}$	$\mathbf{v} = \dfrac{d\mathbf{r}}{dt} = \dot{x}\mathbf{i} + \dot{y}\mathbf{j}$	$\mathbf{a} = \dfrac{d\mathbf{v}}{dt} = \ddot{x}\mathbf{i} + \ddot{y}\mathbf{j}$

	Acceleration	→ Velocity	→ Position
		integrate →	
In one dimension	a	$v = \int a\,dt$	$s = \int v\,dt$
In two dimensions	$\mathbf{a}$	$\mathbf{v} = \int \mathbf{a}\,dt$	$\mathbf{r} = \int \mathbf{v}\,dt$

- Acceleration may be due to change in direction or change in speed or both.
- Using vectors, Newton's second law is $\mathbf{F} = m\mathbf{a}$.

Chapter three

CENTRE OF MASS

Let man then contemplate the whole of nature in her full and grand mystery ... It is an infinite sphere, the centre of which is everywhere, the circumference nowhere.

Blaise Pascal

Figure 3.1, which is drawn to scale, shows a mobile suspended from the point P. The horizontal rods and the strings are light but the geometrically shaped pieces are made of uniform heavy card. Does the mobile balance? If it does, what can you say about the position of its centre of mass?

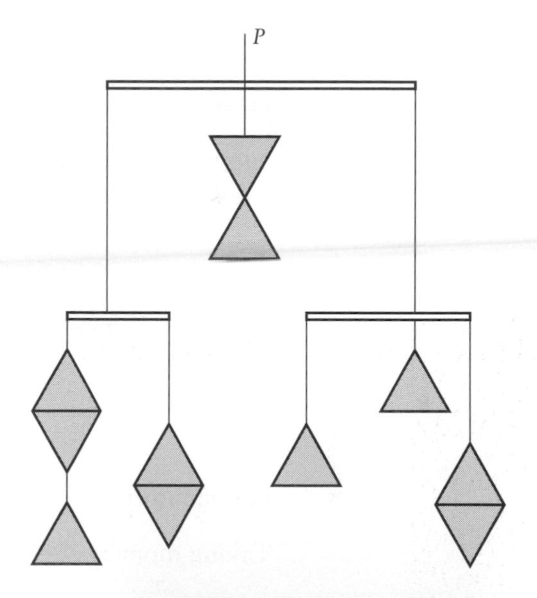

FIGURE 3.1

Where is the centre of mass of the gymnast in the picture (right)?

You have met the concept of centre of mass in the context of two general models.

- *The particle model*
  The centre of mass is the single point at which the whole mass of the body may be taken to be situated.
- *The rigid body model*
  The centre of mass is the balance point of a body with size and shape.

The following examples show how to calculate the position of the centre of mass of a body.

EXAMPLE 3.1

An object consists of three point masses 8 kg, 5 kg and 4 kg attached to a rigid light rod as shown.

FIGURE 3.2

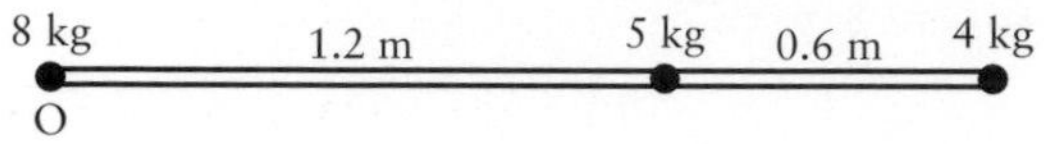

8 kg
1.2 m
5 kg
0.6 m
4 kg
O

Calculate the distance of the centre of mass of the object from end O. (Ignore the mass of the rod.)

Solution Suppose the centre of mass C is $\bar{x}$ m from O. If a pivot were at this position the rod would balance.

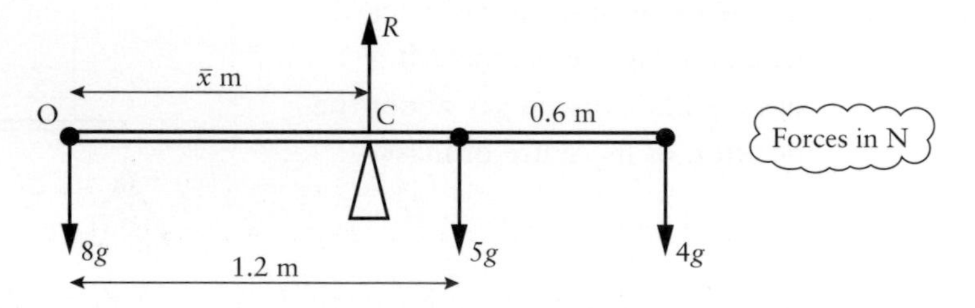

FIGURE 3.3

R
x̄ m
O
C
0.6 m
Forces in N
8g
1.2 m
5g
4g

For equilibrium $\quad R = 8g + 5g + 4g = 17g$

Taking moments of the forces about O gives:

$$\begin{aligned}\text{Total clockwise moment} &= (8g \times 0) + (5g \times 1.2) + (4g \times 1.8)\\ &= 13.2g\,\text{Nm}\\ \text{Total anticlockwise moment} &= R\bar{x}\\ &= 17g\bar{x}\,\text{Nm}\end{aligned}$$

The overall moment must be zero for the rod to be in balance, so

$$17g\bar{x} - 13.2g = 0$$

$$\Rightarrow \quad 17\bar{x} = 13.2$$

$$\Rightarrow \quad \bar{x} = \frac{13.2}{17} = 0.776$$

The centre of mass is 0.776 m from the end O of the rod.

Note that although g was included in the calculation, it cancelled out. The answer depends only on the masses and their distances from the origin and not on the value of g. This leads to the following definition for the position of the centre of mass.

Definition

Consider a set of n point masses $m_1, m_2, \ldots, m_n$ attached to a rigid light rod (whose mass is neglected) at positions $x_1, x_2, \ldots, x_n$ from one end O. The situation is shown in figure 3.4.

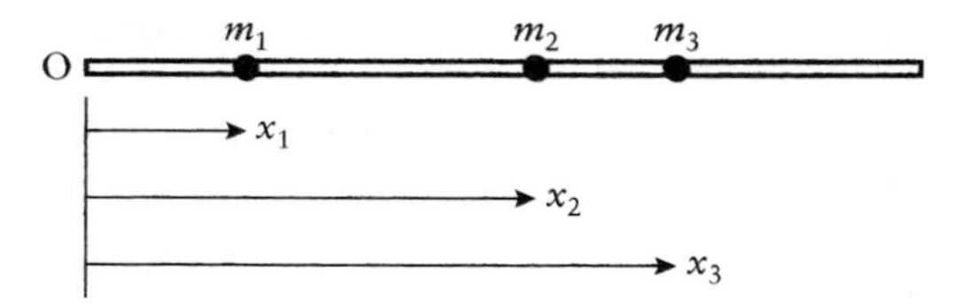

FIGURE 3.4

The position, $\bar{x}$, of the centre of mass relative to O, is defined by the equation

moment of whole mass at centre of mass = sum of moments of individual masses

$$(m_1 + m_2 + m_3 + \ldots)\bar{x} = m_1 x_1 + m_2 x_2 + m_3 x_3 + \ldots$$

or

$$M\bar{x} = \sum_{i=1}^{n} m_i x_i$$

where M is the total mass (or Σm_i).

EXAMPLE 3.2

A uniform rod of length 2 m has mass 5 kg. Masses of 4 kg and 6 kg are fixed at each end of the rod. Find the centre of mass of the rod.

Solution Since the rod is uniform, it can be treated as a point mass at its centre. Figure 3.5 illustrates this situation.

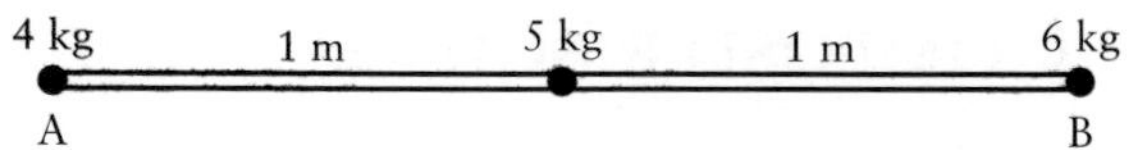

FIGURE 3.5

Taking the end A as origin,

$$\begin{aligned} M\bar{x} &= \Sigma m_i x_i \\ (4 + 5 + 6)\bar{x} &= 4 \times 0 + 5 \times 1 + 6 \times 2 \\ 15\bar{x} &= 17 \\ \bar{x} &= \tfrac{17}{15} \\ &= 1\tfrac{2}{15} \end{aligned}$$

So the centre of mass is 1.133 m from the 4 kg point mass.

Check that the rod would balance about a pivot $1\frac{2}{15}$ m from A.

EXAMPLE 3.3

A rod AB of mass 1.1 kg and length 1.2 m has its centre of mass 0.48 m from the end A. What mass should be attached to the end B to ensure that the centre of mass is at the mid-point of the rod?

Solution Let the extra mass be m kg.

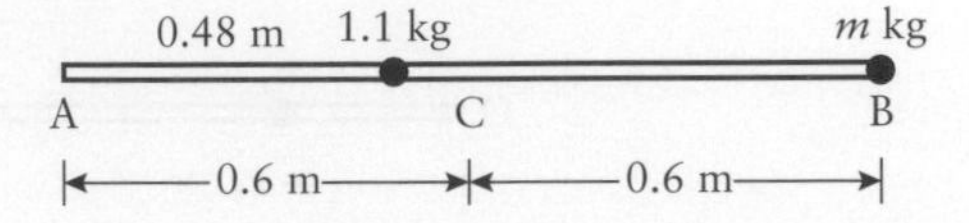

FIGURE 3.6

Method 1

Refer to the mid-point, C, as origin, so $\bar{x} = 0$. Then

$$(1.1 + m) \times 0 = 1.1 \times (-0.12) + m \times 0.6$$

$$\Rightarrow \quad 0.6m = 1.1 \times 0.12$$

$$\Rightarrow \quad m = 0.22$$

The 1.1 mass has negative x referred to C

A mass of 220 grams should be attached to B.

Method 2

Refer to the end A, as origin, so $\bar{x} = 0.6$. Then

$$(1.1 + m) \times 0.6 = 1.1 \times 0.48 + m \times 1.2$$

$$\Rightarrow \quad 0.66 + 0.6m = 0.528 + 1.2m$$

$$\Rightarrow \quad 0.132 = 0.6m$$

$$m = 0.22 \text{ as before}$$

COMPOSITE BODIES

The position of the centre of mass of a composite body such as a cricket bat, tennis racquet or golf club is important to sports people who like to feel its balance. If the body is symmetric then the centre of mass will lie on the axis of symmetry. The next example shows how to model a composite body as a system of point masses so that the methods of the previous section can be used to find the centre of mass.

EXAMPLE 3.4

A squash racquet of mass 200 g and total length 70 cm consists of a handle of mass 150 g whose centre of mass is 20 cm from the end, and a frame of mass 50 g, whose centre of mass is 55 cm from the end.

Find the distance of the centre of mass from the end of the handle.

Solution Figure 3.7 shows the squash racquet and its dimensions.

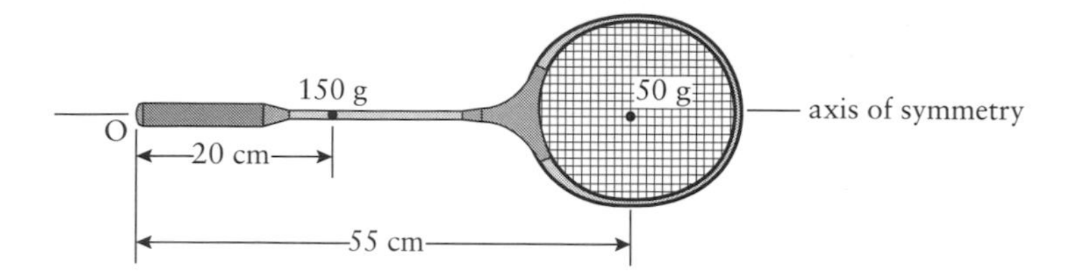

FIGURE 3.7

The centre of mass lies on the axis of symmetry. Model the handle as a point mass of 0.15 kg a distance 0.2 m from O and the frame as a point mass of 0.05 kg a distance 0.55 m from the end O.

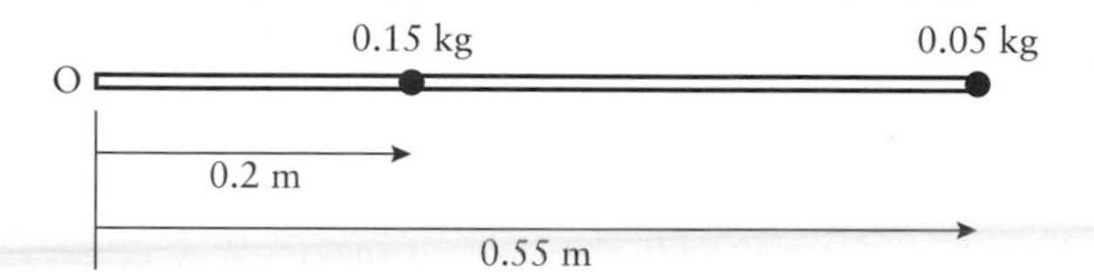

FIGURE 3.8

The distance, $\bar{x}$, of the centre of mass from O is given by

$$(0.15 + 0.05)\bar{x} = (0.15 \times 0.2) + (0.05 \times 0.55)$$
$$\bar{x} = 0.2875$$

The centre of mass of the squash racquet is 28.75 cm from the end of the handle.

Centres of mass for different shapes

If an object has an axis of symmetry, like the squash racquet in the example above, then the centre of mass lies on it.

The table below gives the position of the centre of mass of some uniform objects that you may encounter, or wish to include within models of composite bodies in *Mechanics 2* or *3*.

Body	Position of centre of mass	Diagram
Triangular lamina	$\frac{2}{3}$ along median from vertex	
Solid hemisphere	$\frac{3}{8}r$ from base	r
Hemispherical shell, radius r	$\frac{1}{2}r$ from base	r
Circular arc, radius r, angle at centre 2α	$\frac{r\sin\alpha}{\alpha}$ from centre	2α, r
Sector of circle, radius r, angle at centre 2α	$\frac{2r\sin\alpha}{3\alpha}$ from centre	2α, r
Solid cone or pyramid of height h	$\frac{1}{4}h$ above the base on the line from centre of base to vertex	h, r
Conical shell of height h	$\frac{1}{3}h$ above the base on the line from centre of base to vertex	h, r

EXERCISE 3A

1 The diagrams show point masses attached to rigid light rods. In each case calculate the position of the centre of mass relative to the point O.

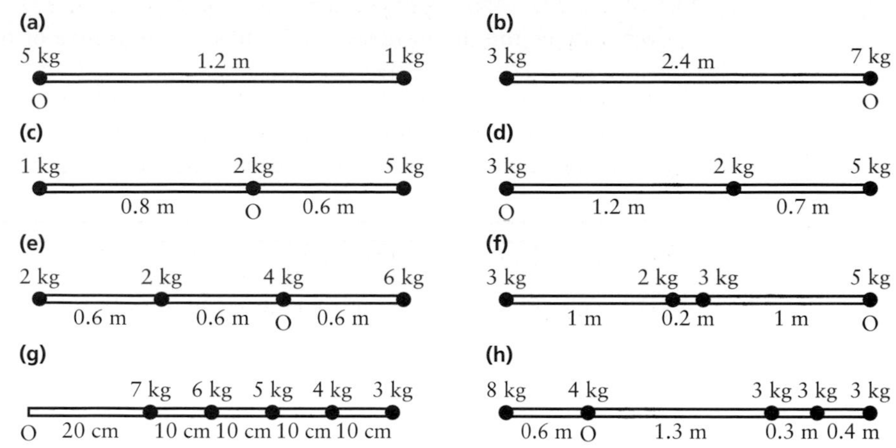

2 A seesaw consists of a uniform plank 4 m long of mass 10 kg. Calculate the centre of mass when two children, of masses 20 kg and 25 kg, sit, one on each end.

3 A weightlifter's bar in a competition has mass 10 kg and length 1 m. By mistake, 50 kg is placed on one end and 60 kg on the other end. How far is the centre of mass of the bar from the centre of the bar itself?

4 The masses of the earth and the moon are 5.98×10^{24} kg and 7.38×10^{22} kg, and the distance between their centres is 3.84×10^{5} km. How far from the centre of the earth is the centre of mass of the earth–moon system?

5 A lollipop lady carries a sign which consists of a uniform rod of length 1.5 m, and mass 1 kg, on top of which is a circular disc of radius 0.25 m and mass 0.2 kg. Find the distance of the centre of mass from the free end of the stick.

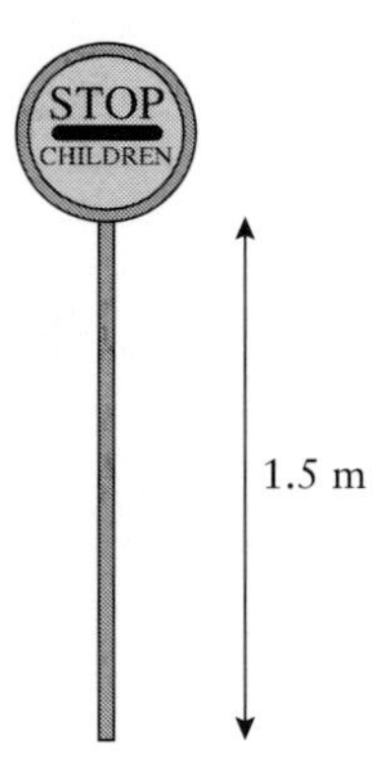

6 A rod has length 2 m and mass 3 kg. The centre of mass should be in the middle but due to a fault in the manufacturing process it is not. This error is corrected by placing a 200 g mass 5 cm from the centre of the rod. Where is the centre of mass of the rod itself?

7 A child's toy consists of four uniform discs, all made out of the same material. They each have thickness 2 cm and their radii are 6 cm, 5 cm, 4 cm and 3 cm. They are placed symmetrically on top of each other to form a tower. How high is the centre of mass of the tower?

8 A standard lamp consists of a uniform heavy metal base of thickness 4 cm, attached to which is a uniform metal rod of length 1.75 m and mass 0.25 kg.

What is the minimum mass for the base if the centre of mass of the lamp is no more than 12 cm from the ground?

9 A uniform scaffold pole of length 5 m has brackets bolted to it as shown in the diagram below. The mass of each bracket is 1 kg.

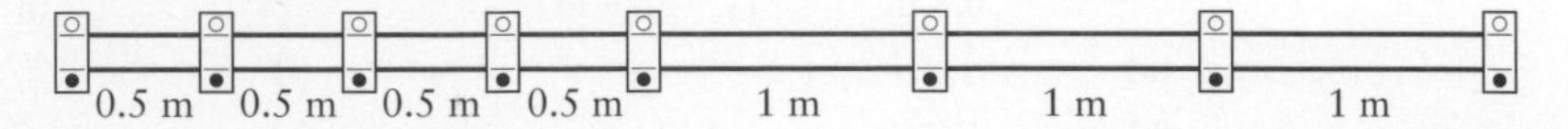

The centre of mass is 2.44 m from the left-hand end. What is the mass of the pole?

10 An object of mass m_1 is placed at one end of a light rod of length l. An object of mass m_2 is placed at the other end. Find the position of the centre of mass.

11 The diagram illustrates a mobile tower crane. It consists of the main vertical section (mass M tonnes), housing the engine, winding gear and controls, and the boom. The centre of mass of the main section is on its centre line. The boom, which has negligible mass, supports the load (L tonnes) and the counterweight (C tonnes). The main section stands on supports at P and Q, distance $2d$ m apart. The counterweight is held at a fixed distance a m from the centre line of the main section and the load at a variable distance l m.

(a) In the case when $C = 3$, $M = 10$, $L = 7$, $a = 8$, $d = 2$ and $l = 13$, find the horizontal position of the centre of mass and say what happens to the crane.

(b) Show that for these values of C, M, a, d and l the crane will not fall over when it has no load, and find the maximum safe load that it can carry.

(c) Formulate two inequalities in terms of C, M, L, a, d and l that must hold if the crane is to be safely loaded or unloaded.

(d) Find, in terms of M, a, d and l, the maximum load that the crane can carry.

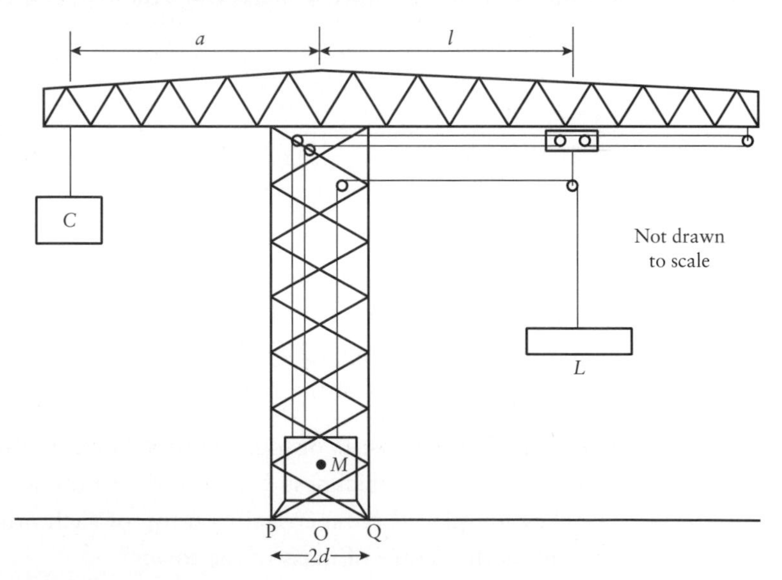

Centre of mass for two-dimensional bodies

The techniques developed for finding the centre of mass using moments can be extended into two dimensions.

If a two-dimensional body consists of a set of n point masses $m_1, m_2, \ldots, m_n$ located at positions $(x_1, y_1), (x_2, y_2), \ldots, (x_n, y_n)$ as in figure 3.9 then the position of the centre of mass of the body $(\bar{x}, \bar{y})$ is given by

$$M\bar{x} = \Sigma m_i x_i \text{ and } M\bar{y} = \Sigma m_i y_i$$

where $M(= \Sigma m_i)$ is the total mass of the body.

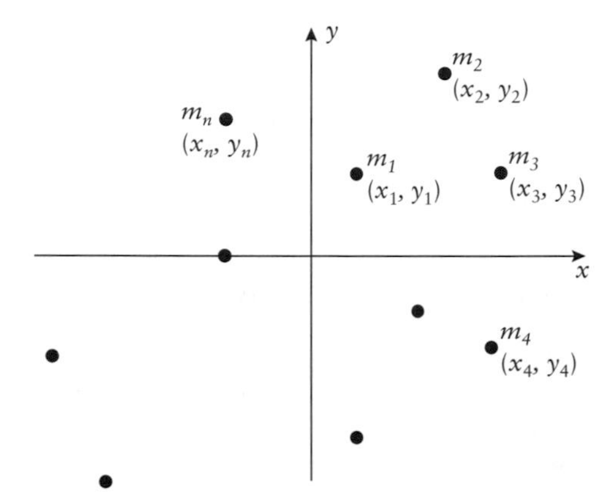

Figure 3.9

The centre of mass of any composite body in two dimensions can be found by replacing each component by a point mass at its centre of mass.

EXAMPLE 3.5

Joanna makes herself a pendant in the shape of a letter J made up of rectangular shapes as shown in figure 3.10.

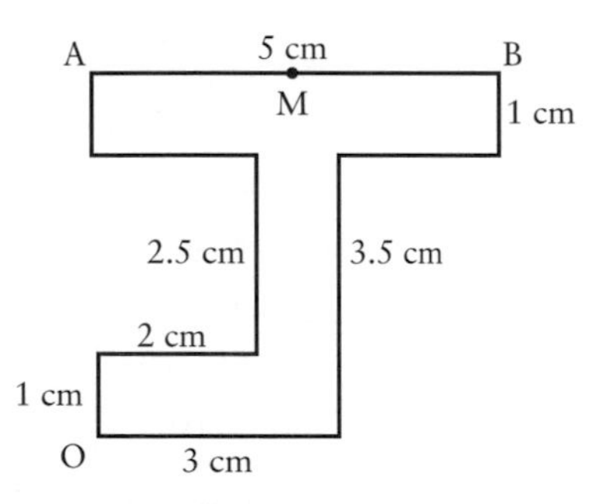

Figure 3.10

(a) Find the position of the centre of mass of the pendant.

(b) Find the angle that AB makes with the horizontal if she hangs the pendant from a point, M, in the middle of AB.

She wishes to hang the pendant so that AB is horizontal.

(c) How far along AB should she place the ring that the suspending chain will pass through?

Solution **(a)** The first step is to split the pendant into three rectangles. The centre of mass of each of these is at its middle, as shown in figure 3.11.

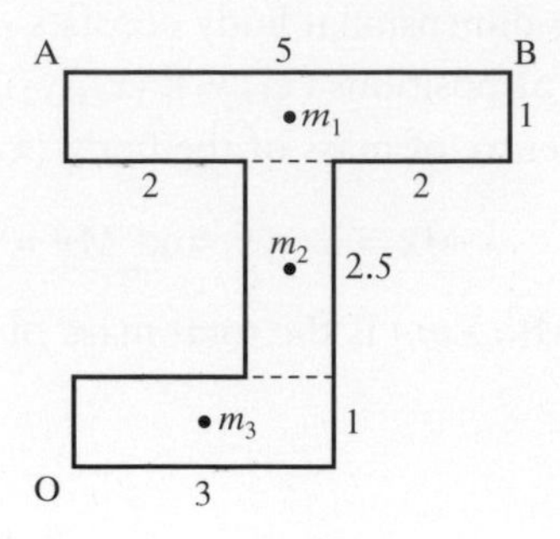

FIGURE 3.11

You can model the pendant as three point masses m_1, m_2 and m_3, which are proportional to the areas of the rectangular shapes. Since the areas are $5\,\text{cm}^2$, $2.5\,\text{cm}^2$ and $3\,\text{cm}^2$, the masses, in suitable units, are 5, 2.5 and 3, and the total mass is $5 + 2.5 + 3 = 10.5$ (in the same units).

The table below gives the mass and position of m_1, m_2 and m_3.

Mass		m_1	m_2	m_3	M
Mass units		5	2.5	3	10.5
Position of	x	2.5	2.5	1.5	$\bar{x}$
centre of mass	y	4	2.25	0.5	$\bar{y}$

Now it is possible to find $\bar{x}$

$$M\bar{x} = \Sigma m_i x_i$$
$$10.5\bar{x} = 5 \times 2.5 + 2.5 \times 2.5 + 3 \times 1.5$$
$$\bar{x} = \frac{23.25}{10.5} = 2.2\,\text{cm}$$

Similarly for $\bar{y}$

$$M\bar{y} = \Sigma m_i y_i$$
$$10.5\bar{y} = 5 \times 4 + 2.5 \times 2.25 + 3 \times 0.5$$
$$\bar{y} = \frac{27.125}{10.5} = 2.6\,\text{cm}$$

The centre of mass is at (2.2, 2.6).

(b) When the pendant is suspended from M, the centre of mass, G, is vertically below M, as shown in figure 3.12.

The pendant hangs like the first diagram but you might find it easier to draw your own diagram like the second.

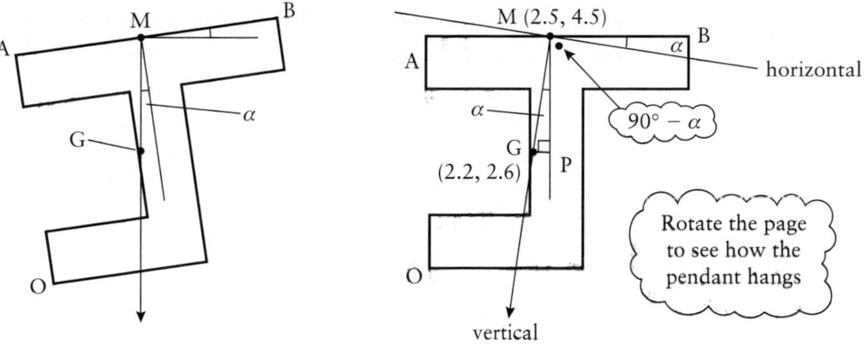

FIGURE 3.12

$$GP = 2.5 - 2.2 = 0.3$$

$$MP = 4.5 - 2.6 = 1.9$$

$$\therefore \quad \tan\alpha = \frac{0.3}{1.9} \Rightarrow \alpha = 9^\circ$$

AB makes an angle of 9° with the horizontal (or 8.5° working with unrounded figures).

(c) For AB to be horizontal the point of suspension must be directly above the centre of mass, and so it is 2.2 cm from A.

EXAMPLE 3.6

Find the centre of mass of a body consisting of a square plate of mass 3 kg and side length 2 m, with small objects of mass 1 kg, 2 kg, 4 kg and 5 kg at the corners of the square.

Solution Figure 3.13 shows the square plate, with the origin taken at the corner at which the 1 kg mass is located. The mass of the plate is represented by a 3 kg point mass at its centre.

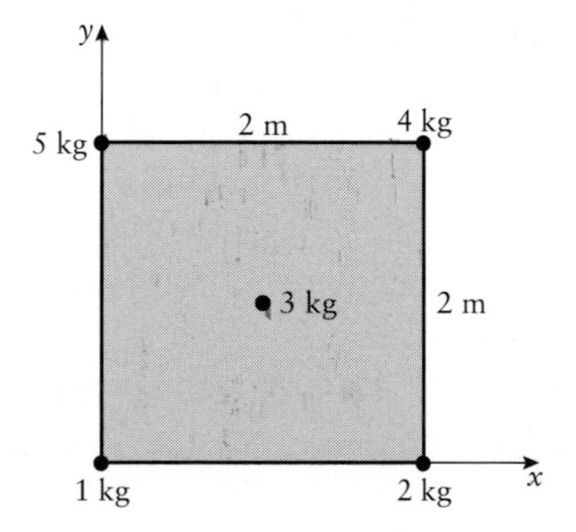

FIGURE 3.13

In this example the total mass M (in kilograms) is 1 + 2 + 4 + 5 + 3 = 15.

$$M\bar{x} = \Sigma m_i x_i$$
$$15\bar{x} = 1 \times 0 + 2 \times 2 + 4 \times 2 + 5 \times 0 + 3 \times 1$$
$$\bar{x} = \tfrac{15}{15} = 1$$

Note the repetition of the masses in the two lines of working

$$M\bar{y} = \Sigma m_i y_i$$
$$15\bar{y} = 1 \times 0 + 2 \times 0 + 4 \times 2 + 5 \times 2 + 3 \times 1$$
$$\bar{y} = \tfrac{21}{15} = 1.4$$

The centre of mass is at the point (1, 1.4).

EXAMPLE 3.7

A metal disc of radius 15 cm has a hole of radius 5 cm cut in it as shown in figure 3.14. Find the centre of mass of the disc.

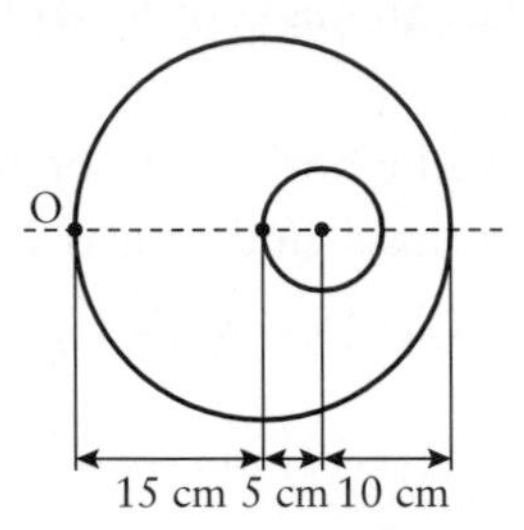

FIGURE 3.14

Solution Think of the original uncut disc as a composite body made up of the final body and a disc to fit into the hole. Since the material is uniform the mass of each part is proportional to its area.

The uncut disc = the final body + the cut out disc

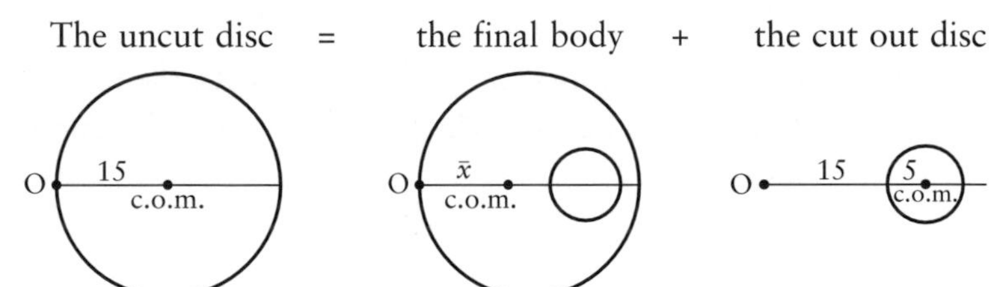

FIGURE 3.15

Area	$15^2\pi = 225\pi$	$15^2\pi - 5^2\pi = 200\pi$	$5^2\pi = 25\pi$
Distance from O to centre of mass	15 cm	$\bar{x}$ cm	20 cm

Taking moments about O

$$225\pi \times 15 = 200\pi \times \bar{x} + 25\pi \times 20$$

$$\Rightarrow \quad \bar{x} = \frac{225 \times 15 - 25 \times 20}{200}$$

$$= 14.375$$

Divide by π

The centre of mass is 14.4 cm from O, that is 0.6 cm to the left of the centre of the disc.

FRAMEWORKS

You can deal with a framework by treating each side as a mass at its centre. If the framework is uniform, the mass is proportional to the length of the side. Note that the centre of mass of a framework is *not* at the same place as that of a lamina of the same shape. The next example shows this.

EXAMPLE 3.8

A uniform wire of length 24 cm is bent into a triangular frame ABC which has a right angle at A. Sides AB, BC and CA are of length 8 cm, 10 cm, and 6 cm respectively.

(a) Find the distance of the centre of mass of the framework from AB and AC.
(b) The framework is hung from a smooth peg at the point C. Find the angle which AC makes with the vertical.

Solution **(a)** The framework is shown in figure 3.16.

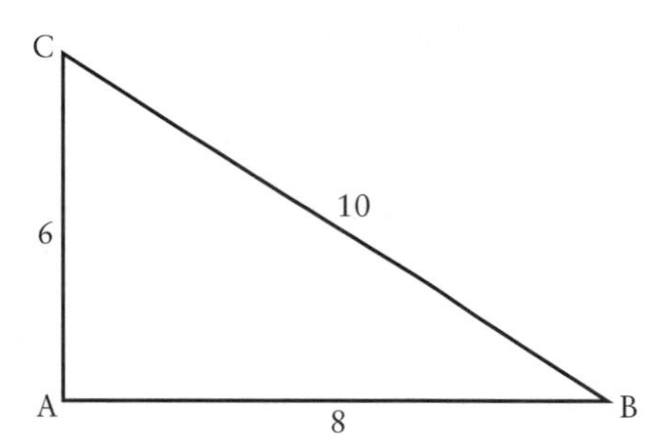

FIGURE 3.16

Taking an origin at A and axes along AB and AC, the centre of mass of AB is at (4, 0), that of BC is at (4, 3), and of AC is at (0, 3).

$$M\bar{x} = \Sigma m_i x_i$$

$$24\bar{x} = 8 \times 4 + 10 \times 4 + 6 \times 0$$

$$\bar{x} = \tfrac{72}{24} = 3$$

$$M\bar{y} = \Sigma m_i y_i$$
$$24\bar{y} = 8 \times 0 + 10 \times 3 + 6 \times 3$$
$$\bar{y} = \tfrac{48}{24} = 2$$

The centre of mass is 2 cm from AB and 3 cm from AC.

(b) When the framework is hung from C, the centre of mass will be directly below C. This is illustrated in figure 3.17.

Since AP = 2 cm, PC = 6 − 2 = 4 cm

$$\tan\alpha = \tfrac{3}{4}$$
$$\alpha = 36.9^\circ$$

The angle between AC and the vertical is 36.9°.

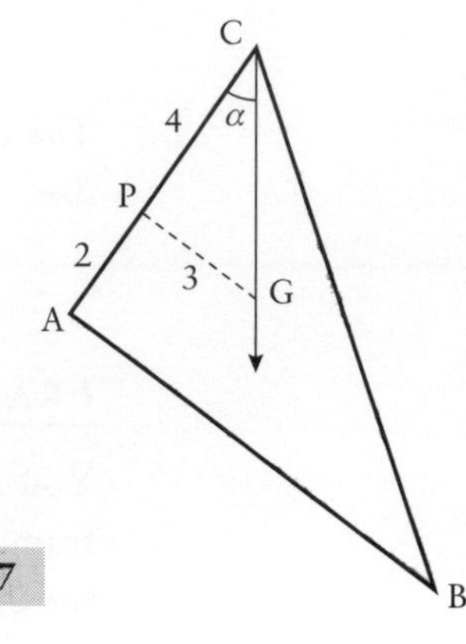

FIGURE 3.17

When an object is about to fall over or topple under gravity alone, the centre of mass must be exactly over the edge of the base. The next example shows how this can be used.

EXAMPLE 3.9

A uniform rectangular plate ABCD with AB = 20 cm and BC = 30 cm is placed on a rough inclined plane with AB in contact with the plane. ABCD lies in a vertical plane and AB is a line of greatest slope of the plane. The plate is on the point of toppling. Find the angle that the plane makes with the horizontal. You may assume that the plane is sufficiently rough to prevent slipping.

Solution Since the plate is uniform, its centre of mass lies at its centre, marked G. As it is on the point of toppling, G is directly over A, as shown in figure 3.18.

From the diagram $\tan\alpha = \tfrac{10}{15}$

$$\alpha = 33.7^\circ$$

The plane is at 33.7° to the horizontal.

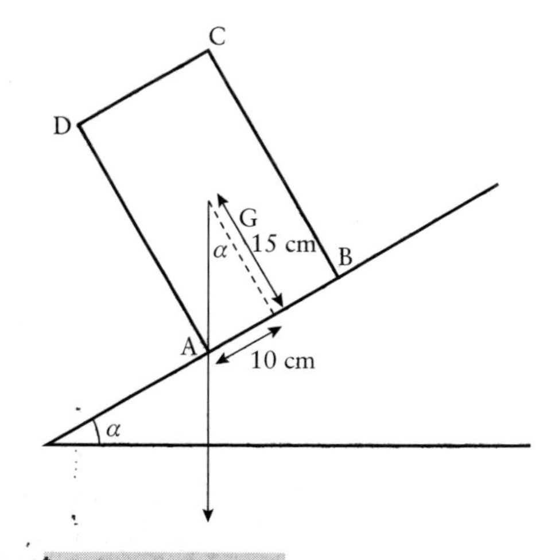

FIGURE 3.18

EXERCISE 3B

1 Find the centre of mass of the following sets of point masses.

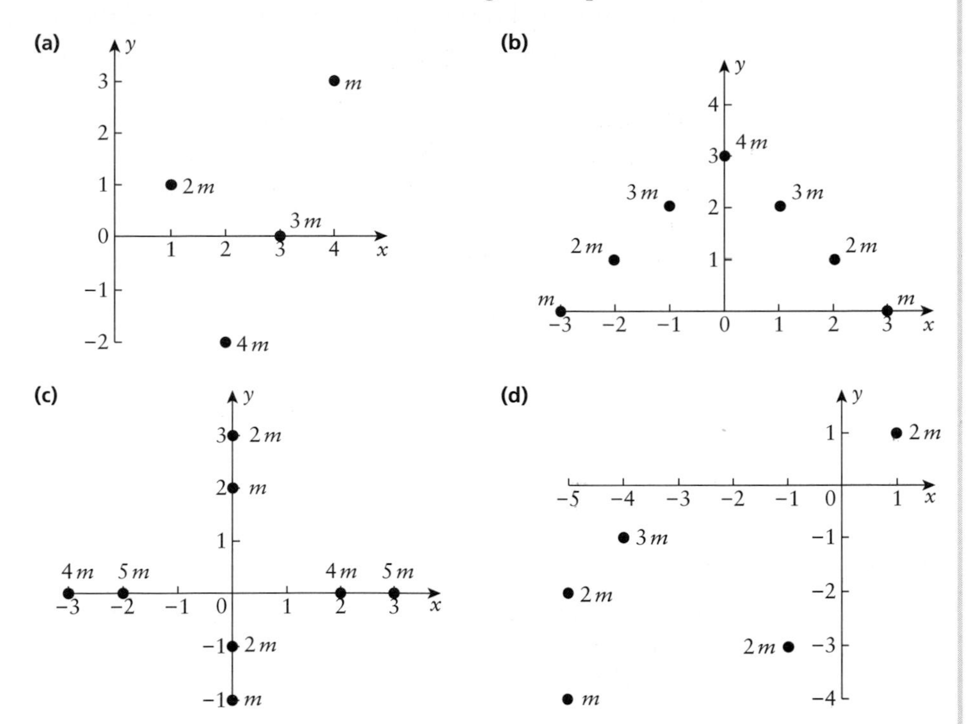

2 Masses of 1, 2, 3 and 4 grams are placed at the corners A, B, C and D of a square piece of uniform cardboard of side 10 cm and mass 5 g. Find the position of the centre of mass relative to axes through AB and AD.

3 As part of a Christmas lights display, letters are produced by mounting bulbs in holders 30 cm apart on light wire frames. The combined mass of a bulb and its holder is 200 g. Find the position of the centre of mass for each of the letters shown below, in terms of its horizontal and vertical displacement from the bottom left-hand corner of the letter.

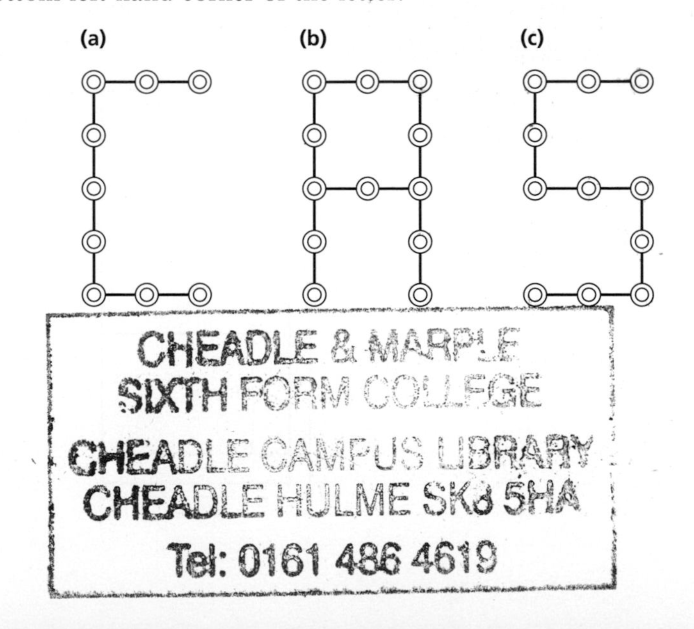

4 Four people of masses 60 kg, 65 kg, 62 kg and 75 kg sit on the four seats of the fairground ride shown below. The seats and the connecting arms are light. Find the radius of the circle described by the centre of mass when the ride rotates about O.

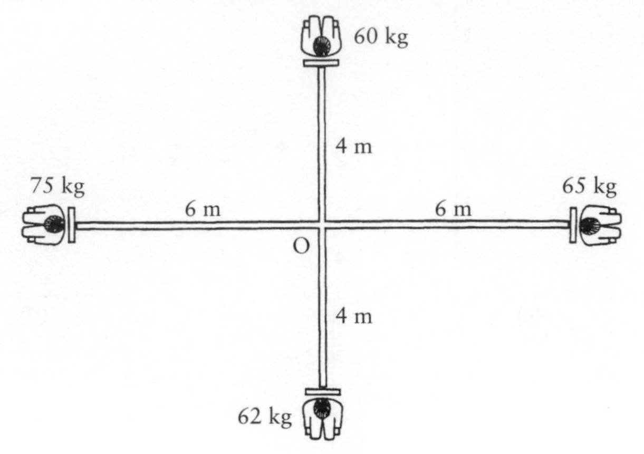

5 The following shapes are made out of uniform card. For each shape find the coordinates of the centre of mass relative to O.

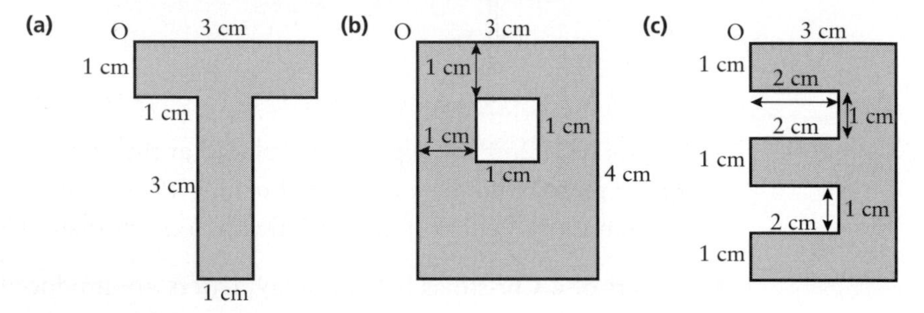

6 A filing cabinet has the dimensions shown in the diagram. The body of the cabinet has mass 20 kg and its construction is such that its centre of mass is at a height of 60 cm, and is 25 cm from the back of the cabinet. The mass of a drawer and its contents may be taken to be 10 kg and its centre of mass to be 10 cm above its base and 30 cm from its front face.

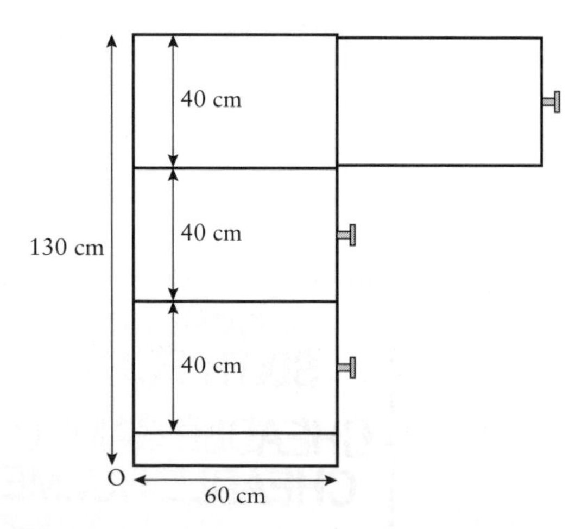

(a) Find the position of the centre of mass when all the drawers are closed.
(b) Find the position of the centre of mass when the top two drawers are fully open.
(c) Show that when all three drawers are fully opened the filing cabinet will tip over.
(d) Two drawers are fully open. How far can the third one be opened without the cabinet tipping over?

7 Uniform wooden bricks have length 20 cm and height 5 cm. They are glued together as shown in the diagram with each brick 5 cm to the right of the one below it. The origin is taken to be at O.

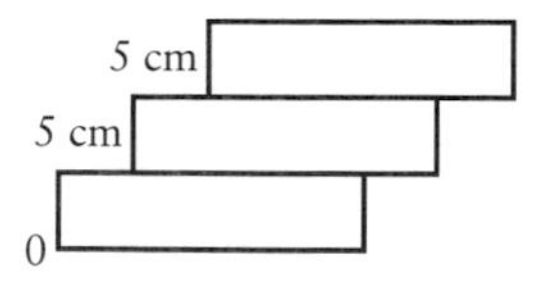

(a) Find the coordinates of the centre of mass for
(i) 1 (ii) 2 (iii) 3 (iv) 4 (v) 5 bricks.
(b) How many bricks is it possible to assemble in this way without them tipping over?
(c) If the displacement is changed from 5 cm to 2 cm find the coordinates of the centre of mass for n bricks. How many bricks can now be assembled?
(d) If the displacement is $\frac{1}{2}$ cm, what is the maximum height possible for the centre of mass of such an assembly of bricks without them tipping over?

8 A pendant is made from a uniform circular disc of mass $4m$ and radius 2 cm with a decorative edging of mass m as shown. The centre of mass of the decorative edging is 1 cm below the centre, O, of the disc. The pendant is symmetrical about the diameter AB.

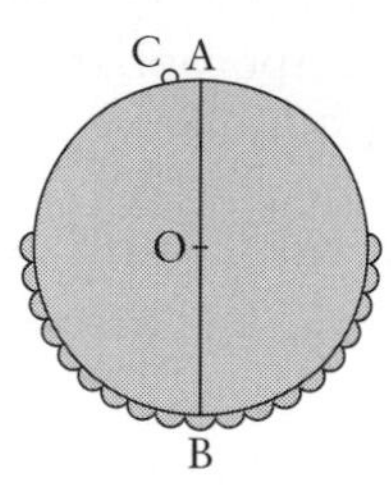

(a) Find the position of the centre of mass of the pendant.

The pendant should be hung from A but the light ring for hanging it is attached at C where angle AOC is 10°.

(b) Find the angle between AB and the vertical when the pendant is hung from C.

9 A uniform rectangular lamina, ABCD, where AB is of length a and BC of length $2a$, has a mass $10m$. Further point masses m, $2m$, $3m$ and $4m$ are fixed to the points A, B, C and D, respectively.

(a) Find the centre of mass of the system relative to x and y axes along AB and AD respectively.

(b) If the lamina is suspended from the point A find the angle that the diagonal AC makes with the vertical.

(c) To what must the mass at point D be altered if this diagonal is to hang vertically?

[MEI]

10

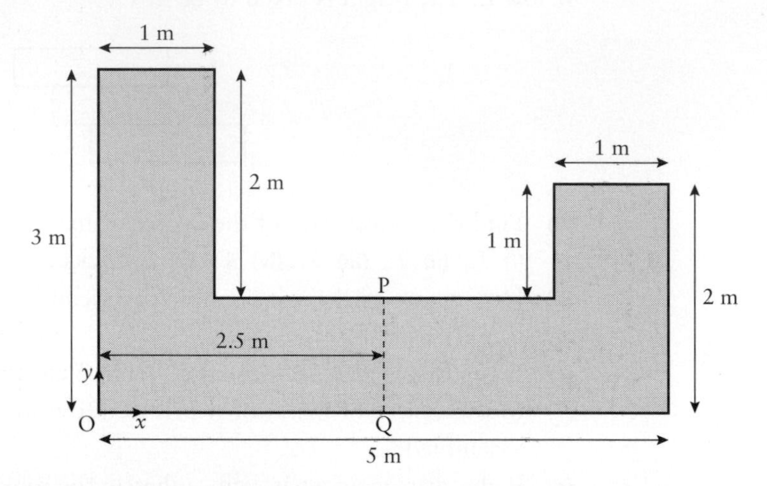

The diagram gives the dimensions of the design of a uniform metal plate. Using a coordinate system with O as origin, the x and y axes as shown and 1 metre as 1 unit,

(a) show that the centre of mass has y coordinate 1 and find its x coordinate.

The design requires the plate to have its centre of mass half-way across (i.e. on the line PQ in the diagram), and in order to achieve this a circular hole centred on $(\frac{1}{2}, \frac{1}{2})$ is considered.

(b) Find the appropriate radius for such a hole and explain why this idea is not feasible.

It is then decided to cut two circular holes each of radius r, both centred on the line $x = \frac{1}{2}$. The first hole is centred at $(\frac{1}{2}, \frac{1}{2})$ and the centre of mass of the plate is to be at P.

(c) Find the value of r and the coordinates of the centre of the second hole.

[MEI]

EXERCISE 3C **Examination-style questions**

1 A uniform lamina is in the shape of a rectangle ABCD with a semi-circle attached with its diameter along BC. AB = 12 cm and BC = 8 cm.

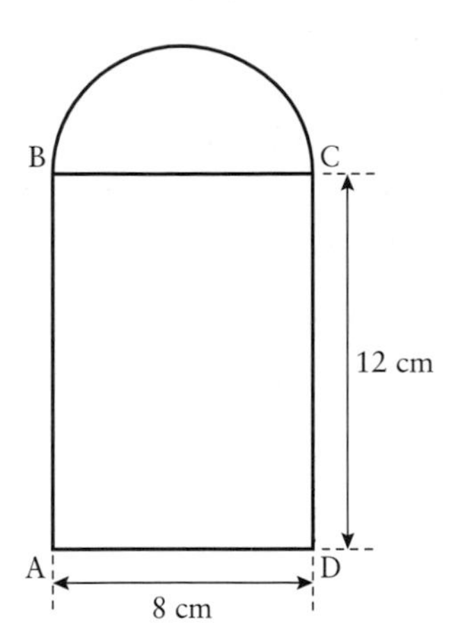

(a) Find the distance of the centre of mass of the lamina from AD.

The lamina is placed on a rough inclined plane with AD on a line of greatest slope. The angle the plane makes with the horizontal is slowly increased.

(b) Find, in degrees to 1 decimal place, the angle the plane makes with the horizontal as the lamina topples.

2 A modeller is making a part for the front of a model steam train from uniform plate. The rectangle ABCD has AB = 15 cm and AD = 20 cm, and two circular holes have been cut. Each has radius 4 cm, and they are symmetrically placed with their centres 5 cm from AD and 5 cm from the edges AB and CD, as shown in the diagram.

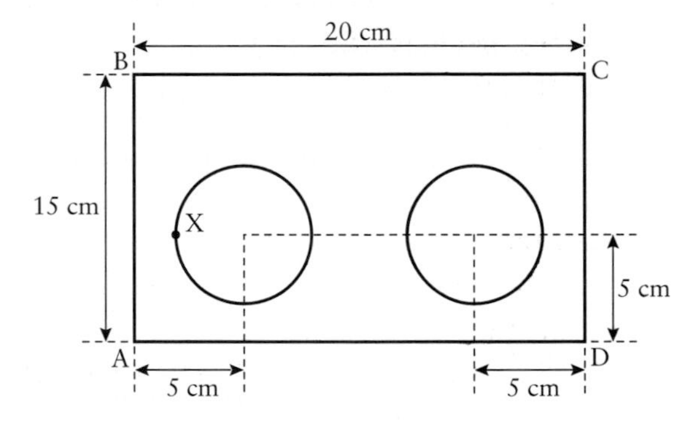

(a) Find the distance of the centre of mass from AD.

The part is hung from a small peg at X.

(b) Find, to the nearest degree, the angle that AB makes with the horizontal.

3 A pointer is made from a uniform lamina and consists of a square of side 10 cm. At one end a uniform semi-circle of the same material and radius 5 cm is attached, and at the other an isosceles triangle of base 10 cm and height 5 cm is removed. The pointer is shown in the diagram.

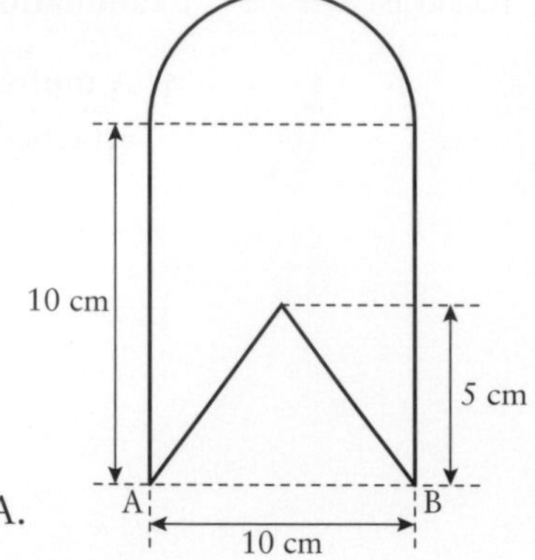

(a) Find the position of the centre of mass of the pointer.

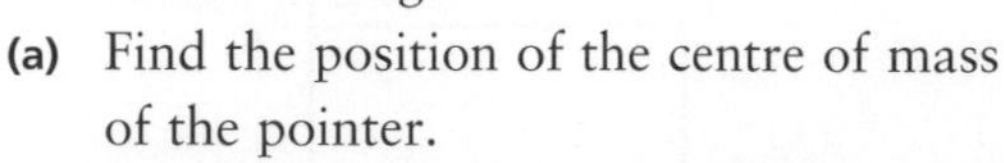

The pointer is smoothly suspended from point A.

(b) Find, to the nearest degree, the angle AB makes with the horizontal.

4 A uniform rectangular lamina ABCD, with AD = a and AB = b, of mass $5m$ has three point masses $2m$, $4m$ and $3m$ attached at A, C and D respectively. Find, in terms of a and b, the distance of the centre of mass of the loaded lamina

(a) from AB

(b) from AD.

The loaded lamina is now freely suspended from B. BA makes an angle of α with the downward vertical, where $\tan\alpha = \frac{3}{4}$.

(c) Show that $76a = 45b$.

5 A uniform rectangular lamina ABCD, where AB = a and AD = $2a$, has mass M. Two particles of mass kM and $3kM$ are attached to B and D respectively.

(a) Find the distances of the centre of mass of the loaded lamina from AB and AD in terms of k and a.

(b) The lamina is now freely suspended from A, and AD makes an angle of α with the downward vertical, where $\tan\alpha = \frac{7}{38}$. Find k.

6 A decoration is made from uniform card in the shape of a letter L as shown in the diagram.

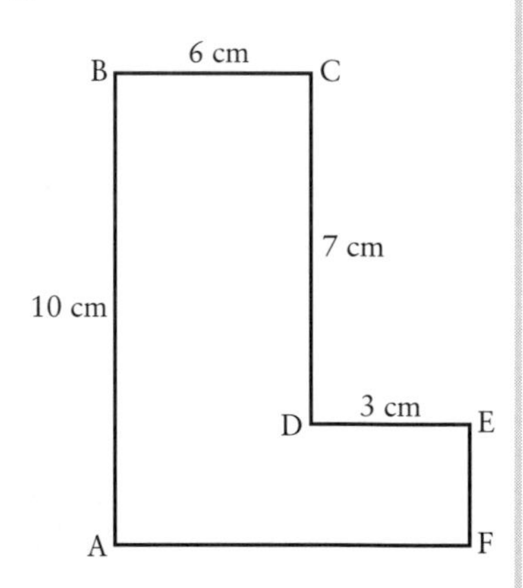

Find the distance of the centre of mass

(a) from AB

(b) from AF.

The decoration is freely suspended from point C.

(c) Find, in degrees to 1 decimal place, the angle which CD makes with the vertical.

The decoration has mass m. In order to make it hang with CD vertical, a particle of mass km is attached at F.

(d) Find k.

7 A set square is made from uniform plastic, and is in the shape of a right-angled triangle OAB, with a right angle at O, and angles of 60° at A and 30° at B. The hypotenuse AB is of length 15 cm. A triangle PQR is cut from the centre. PQ is of length 2.6 cm, and is parallel to OA and 1.8 cm away from OA. PR is of length 4.5 cm and is parallel to OB and 1.8 cm away from OB. The set square is illustrated in the diagram.

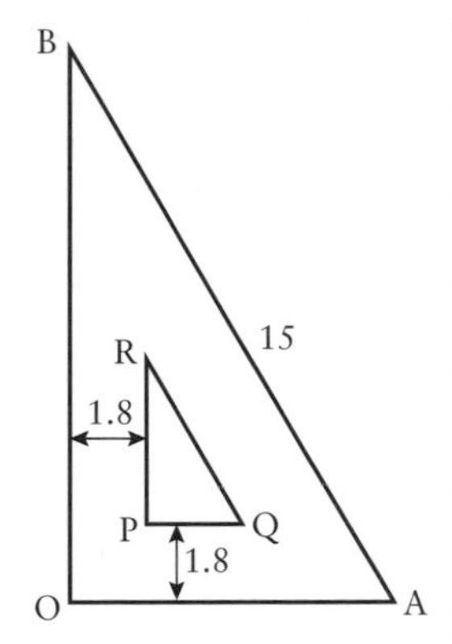

Find the distance of the centre of mass

(a) from OB

(b) from OA.

The set square is hung on a smooth peg at R.

(c) Find, in degrees to 1 decimal place, the acute angle which OB makes with the vertical.

8 A uniform rectangular lamina ABCD is of mass $7m$. The side AD is of length a and the side AB is of length b. Point masses $2m$, $5m$, $3m$ and $4m$ are attached to A, B, C and D respectively.

(a) Find, in terms of a and b, the distance of the centre of mass of the loaded lamina from AB and AD.

The loaded lamina is freely suspended from the mid-point of AB.

(b) Find, in terms of a and b, the angle that AB makes with the horizontal.

An extra mass is now attached to the mid-point of AD to make the loaded lamina hang with AB horizontal.

(c) Find the extra mass in terms of m.

9 A coat hanger is constructed from uniform wire and has the dimensions shown in the diagram.

AB and BC are 26 cm, AC is 48 cm, BD is 8 cm. The arc DEF may be treated as a semi-circle with the diameter DF being horizontal and the radius 4 cm long. Find the distance of the centre of mass

(a) from the vertical line OB

(b) from AC.

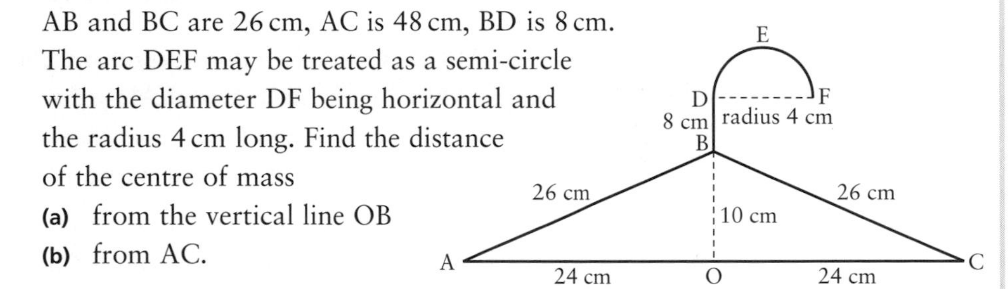

The coat hanger is hung from a small peg at E, which is the point half-way round the semi-circle.

(c) Find, in degrees to 3 significant figures, the angle that AC makes with the horizontal.

(d) Explain why the peg must be rough.

10 A uniform wire of mass m is bent to form the pentagonal frame shown in the diagram.

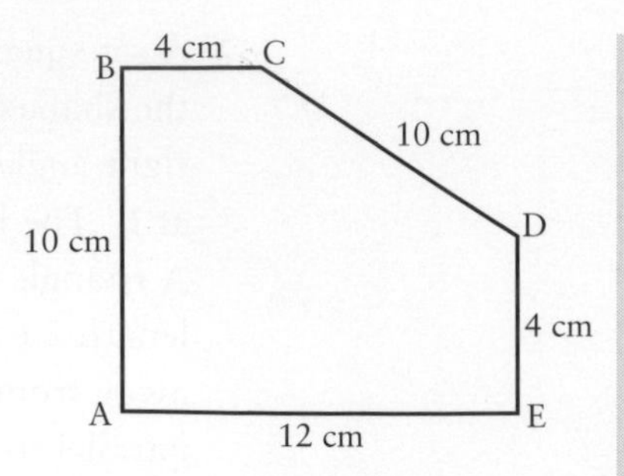

Find the distance of the centre of mass

(a) from AB

(b) from AE.

A particle of mass km is attached at A and the loaded frame is freely suspended from B.

(c) Explain why the line AB cannot be vertical, whatever the value of k.

The loaded frame is now freely suspended from C and the line AB is found to be vertical.

(d) Find k.

11 Three particles of mass $3m$, $5m$ and λm are placed at points with coordinates $(4, 0)$, $(0, -3)$ and $(4, 2)$ respectively. The centre of mass of the system of three particles is at $(2, k)$.

(a) Show that $\lambda = 2$.

(b) Calculate the value of k.

[Edexcel]

12

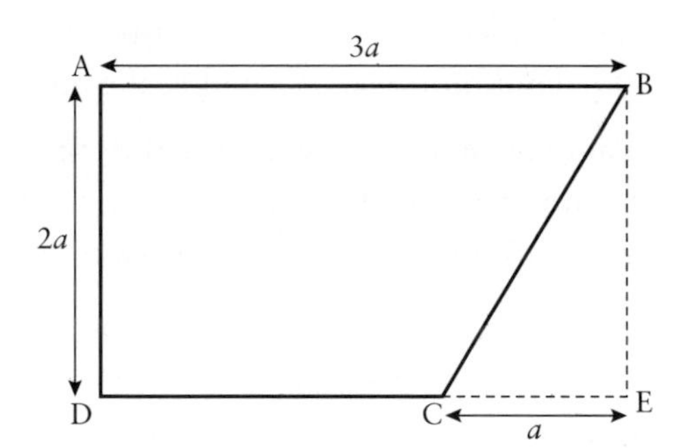

A uniform lamina ABCD is made by taking a uniform sheet of metal in the form of a rectangle ABED, with AB = $3a$ and AD = $2a$, and removing the triangle BCE, where C lies on DE and CE = a, as shown in the diagram.

(a) Find the distance of the centre of mass of the lamina from AD.

The lamina has mass M. A particle of mass m is attached to the lamina at B. When the loaded lamina is freely suspended from the mid-point of AB, it hangs in equilibrium with AB horizontal.

(b) find m in terms of M.

[Edexcel]

13

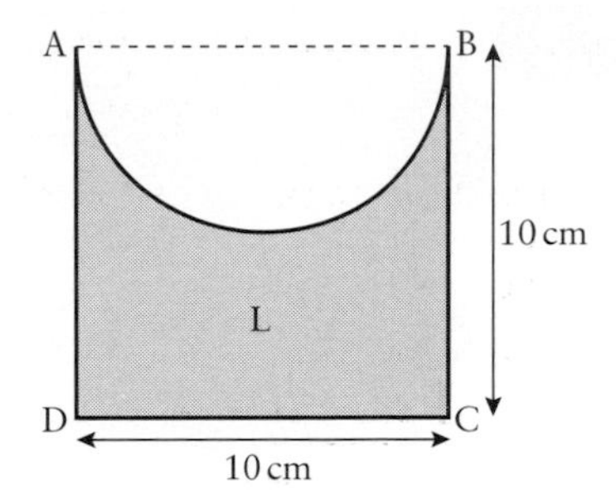

A uniform lamina L is formed by taking a uniform square sheet of material ABCD, of side 10 cm, and removing the semi-circle with diameter AB from the square, as shown in the diagram.

(a) Find, in cm to 2 decimal places, the distance of the centre of mass of the lamina L from the mid-point of AB.

[The centre of mass of a uniform semi-circular lamina, radius a, is at a distance $\frac{4a}{3\pi}$ from the centre of the bounding diameter.]

The lamina is freely suspended from D and hangs at rest.

(b) Find, in degrees to one decimal place, the acute angle between CD and the vertical.

[Edexcel, adapted]

KEY POINTS

1 The centre of mass of a body has the property that *the moment, about any point, of the whole mass of the body taken at the centre of mass is equal to the sum of the moments of the various particles comprising the body.*

$$M\bar{\mathbf{r}} = \Sigma m_i \mathbf{r}_i \qquad \text{where } M = \Sigma m_i$$

2 In one dimension

$$M\bar{x} = \Sigma m_i x_i$$

3 In two dimensions

$$M\bar{x} = \Sigma m_i x_i \quad \text{and} \quad M\bar{y} = \Sigma m_i y_i$$

4 When a body is suspended from a single point, the centre of mass is directly below that point.

5 When a body is on the point of toppling, the centre of mass is directly over the edge of the base.

Chapter four

ENERGY, WORK AND POWER

I like work: it fascinates me. I can sit and look at it for hours.

Jerome K. Jerome

M C Escher's 'Waterfall'

This is a picture of a perpetual motion machine. What does this term mean and will this one work?

ENERGY AND MOMENTUM

When describing the motion of objects in everyday language the words *energy* and *momentum* are often used quite loosely and sometimes no distinction is made between them. In mechanics they must be defined precisely.

For an object of mass m moving with velocity $\mathbf{v}$

Kinetic energy = $\frac{1}{2}mv^2$ (this is the energy it has due to its motion)

Momentum = $m\mathbf{v}$

Notice that kinetic energy is a scalar quantity with magnitude only, but momentum is a vector in the same direction as the velocity.

Both the kinetic energy and the momentum are liable to change when a force acts on a body and you will learn more about how the energy is changed in this chapter. You met one-dimensional momentum in *Mechanics 1* and will meet momentum again in Chapter 5.

WORK AND ENERGY

In everyday life you encounter many forms of energy such as heat, light, electricity and sound. You are familiar with the conversion of one form of energy to another: from chemical energy stored in wood to heat energy when you burn it, from electrical energy to the energy of a train's motion, and so on. The S.I. unit for energy is the joule, J.

MECHANICAL ENERGY AND WORK

In mechanics two forms of energy are particularly important.

Kinetic energy is the energy which a body possesses because of its motion.

> *The kinetic energy of a moving object* = $\frac{1}{2}$ × *mass* × *(speed)*^{2} .

Potential energy is the energy which a body possesses because of its position. It may be thought of as stored energy which can be converted into kinetic or other forms of energy. You will meet this again, later in this chapter.

The energy of an object is usually changed when it is acted on by a force. When a force is applied to an object which moves in the direction of its line of action, the force is said to do *work*. For a constant force this is defined as follows.

> *The work done by a constant force = force × distance moved in the direction of the force.*

The following examples illustrate how to use these ideas.

EXAMPLE 4.1

A brick, initially at rest, is raised by a force averaging 40 N to a height 5 m above the ground where it is left stationary. How much work is done by the force?

Solution The work done by the force raising the brick is

$$40 \times 5 = 200\,\text{J}$$

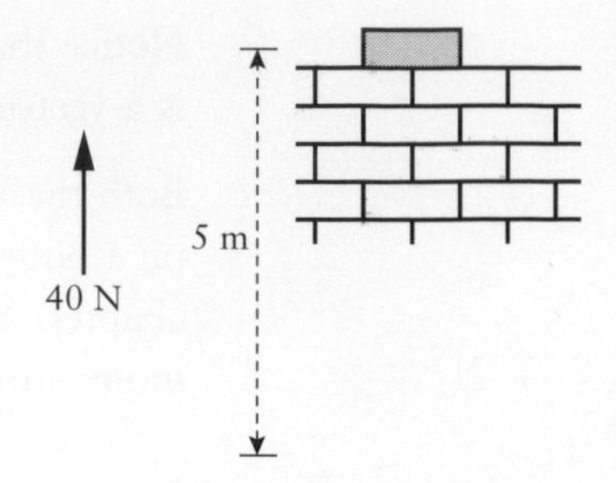

Figure 4.1

Examples 4.2 and 4.3 show how the work done by a force can be related to the change in kinetic energy of an object.

EXAMPLE 4.2

A train travelling on level ground is subject to a resisting force (from the brakes and air resistance) of 250 kN for a distance of 5 km. How much kinetic energy does the train lose?

Solution The forward force is −250 000 N.

Work and energy have the same units

The work done by it is $-250\,000 \times 5000 = -1\,250\,000\,000$ J.

Hence $-1\,250\,000\,000$ J of kinetic energy are gained by the train, in other words $+1\,250\,000\,000$ J of kinetic energy are lost and the train slows down. This energy is converted to other forms such as heat and perhaps a little sound.

EXAMPLE 4.3

A car of mass m kg is travelling at $u\,\text{m s}^{-1}$ when the driver applies a constant driving force of F N. The ground is level and the road is straight and air resistance can be ignored. The driving force will have an effect on the speed of the car. Suppose it increases to $v\,\text{m s}^{-1}$ in a period of t s over a distance of s m.

Treating the car as a particle and applying Newton's second law

$$F = ma$$

$$a = \frac{F}{m}$$

Since F is assumed constant, the acceleration is constant also so using $v^2 = u^2 + 2as$

$$v^2 = u^2 + \frac{2Fs}{m}$$

$$\Rightarrow \quad \tfrac{1}{2}mv^2 = \tfrac{1}{2}mu^2 + Fs$$

$$Fs = \tfrac{1}{2}mv^2 - \tfrac{1}{2}mu^2$$

Thus *work done by force = final kinetic energy – initial kinetic energy* of car.

THE WORK–ENERGY PRINCIPLE

Examples 4.4 and 4.5 illustrate the *work–energy principle* which states that

> *The total work done by the forces acting on a body is equal to the increase in the kinetic energy of the body.*

EXAMPLE 4.4

A sledge of total mass 30 kg, initially moving at 2 ms^{-1}, is pulled 14 m across smooth horizontal ice by a horizontal rope in which there is a constant tension of 45 N. Find its final velocity.

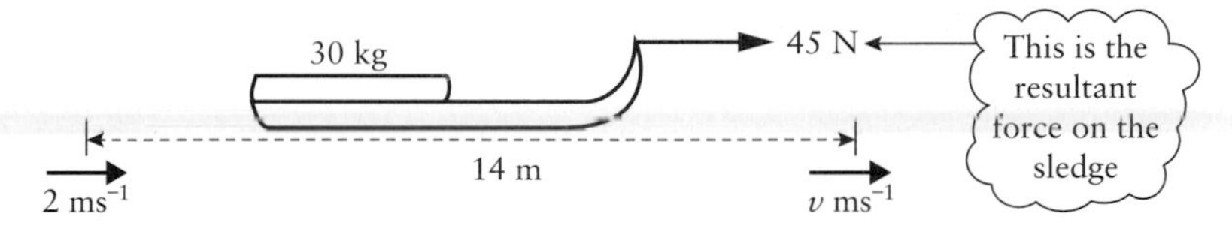

FIGURE 4.2

Solution Since the ice is smooth, the work done by the force is all converted into kinetic energy and the final velocity can be found using

work done by the force = final kinetic energy – initial kinetic energy

$$45 \times 14 = \tfrac{1}{2} \times 30 \times v^2 - \tfrac{1}{2} \times 30 \times 2^2$$

Giving $v^2 = 46$ and the final velocity of the sledge as 6.8 ms^{-1}.

EXAMPLE 4.5

The combined mass of a cyclist and her bicycle is 65 kg. She accelerated from rest to 8 ms^{-1} in 80 m along a horizontal road.

(a) Calculate the work done by the net force in accelerating the cyclist and her bicycle.

(b) Hence calculate the net forward force (assuming the force to be constant).

Solution

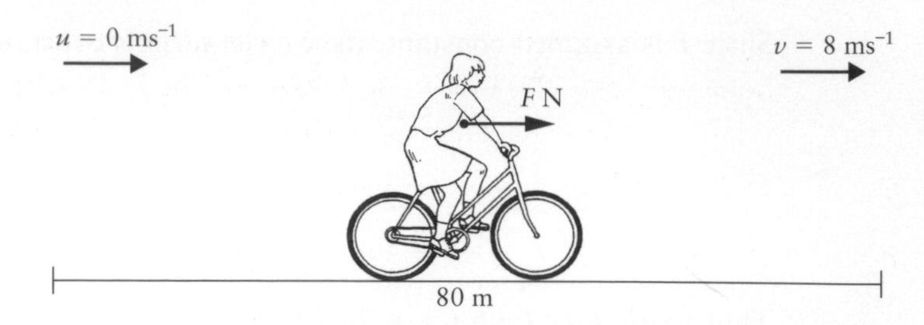

FIGURE 4.3

(a) The work done by the net force F is given by

$$\begin{aligned} \text{work} &= \text{final K.E.} - \text{initial K.E.} \\ &= \tfrac{1}{2}mv^2 - \tfrac{1}{2}mu^2 \\ &= \tfrac{1}{2} \times 65 \times 8^2 - 0 \\ &= 2080\,\text{J} \end{aligned}$$

The work done is 2080 J.

(b)
$$\begin{aligned} \text{Work done} &= Fs \\ &= F \times 80 \\ \text{So} \quad 80F &= 2080 \\ F &= 26 \end{aligned}$$

The net forward force is 26 N.

WORK

It is important to realise that

- work is done by a force
- work is only done when there is movement
- a force only does work on an object when it has a component in the direction of motion of the object.

It is quite common to speak of the work done by a person, say in pushing a lawn mower. In fact this is the work done by the force of the person on the lawn mower.

Notice that if you stand holding a brick stationary above your head, painful though it may be, the force you are exerting on it is doing no work. Nor is this vertical force doing any work if you walk round the room keeping the brick at the same height. However, once you start climbing the stairs, a component of the brick's movement is in the direction of the upward force that you are exerting on it, so the force is now doing some work.

When applying the work–energy principle, you have to be careful to include *all* the forces acting on the body. In the example of a brick of weight 40 N being raised 5 m vertically, starting and ending at rest, the change in kinetic energy is clearly zero.

This seems paradoxical when it is clear that the force which raised the brick has done $40 \times 5 = 200$ J of work. However, the brick was subject to another force, namely its weight, which did $-40 \times 5 = -200$ J of work on it, giving a total of $200 + (-200) = 0$ J.

CONSERVATION OF MECHANICAL ENERGY

The net forward force on the cyclist in Example 4.5 is the girl's driving force minus resistive forces such as air resistance and friction in the bearings. In the absence of such resistive forces, she would gain more kinetic energy; also the work she does against them is lost. It is dissipated as heat and sound. Contrast this with the work a cyclist does against gravity when going uphill. This work can be recovered as kinetic energy on a downhill run. The work done against the force of gravity is conserved and gives the cyclist potential energy (see page 84).

Forces such as friction which result in the dissipation of mechanical energy are called *dissipative forces*. Forces which conserve mechanical energy are called *conservative forces*. The force of gravity is a conservative force and so is the tension in an elastic string; you can test this using an elastic band.

EXAMPLE 4.6

A bullet of mass 25 g is fired at a wooden barrier 3 cm thick. When it hits the barrier it is travelling at $200\,\text{ms}^{-1}$. The barrier exerts a constant resistive force of 5000 N on the bullet.

(a) Does the bullet pass through the barrier and if so with what speed does it emerge?

(b) Is energy conserved in this situation?

Solution

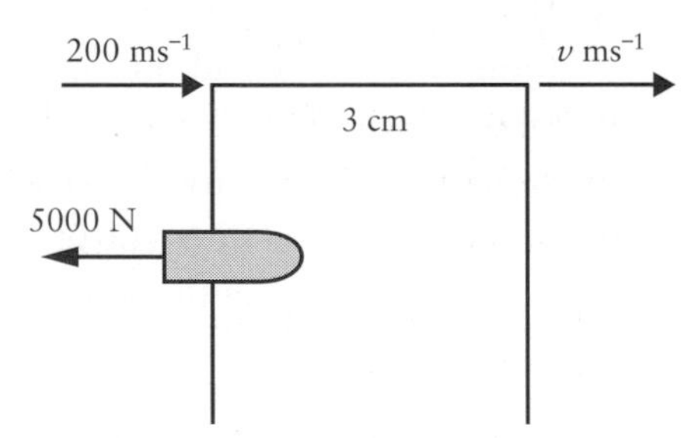

FIGURE 4.4

(a) The work done *by* the force is defined as the product of the force and the distance moved *in the direction of the force*. Since the bullet is moving in the direction opposite to the net resistive force, the work done by this force is negative.

$$\text{Work done} = -5000 \times 0.03\text{ J}$$
$$= -150\text{ J}$$

The initial kinetic energy of the bullet is

$$\begin{aligned}\text{Initial K.E.} &= \tfrac{1}{2}mu^2\\ &= \tfrac{1}{2} \times 0.025 \times 200^2\\ &= 500\,\text{J}\end{aligned}$$

A loss in energy of 150 J will not reduce kinetic energy to zero, so the bullet will still be moving on exit.

Since the work done is equal to the change in kinetic energy,

$$-150 = \tfrac{1}{2}mv^2 - 500$$

Solving for v

$$\begin{aligned}\tfrac{1}{2}mv^2 &= 500 - 150\\ v^2 &= \frac{2 \times (500 - 150)}{0.025}\\ v &= 167 \text{ (to nearest whole number)}\end{aligned}$$

So the bullet emerges from the barrier with a speed of $167\,\text{ms}^{-1}$.

(b) Total energy is conserved but there is a loss of mechanical energy of $\frac{1}{2}mu^2 - \frac{1}{2}mv^2 = 150\,\text{J}$. This energy is converted into non-mechanical forms such as heat and sound.

EXAMPLE 4.7

An aircraft of mass m kg is flying at a constant velocity $v\,\text{ms}^{-1}$ horizontally. Its engines are providing a horizontal driving force F N.

(a) Draw a diagram showing the driving force, the lift force L N, the air resistance (drag force) R N and the weight of the aircraft.

(b) State which of these forces are equal in magnitude.

(c) State which of the forces are doing no work.

(d) In the case when $m = 100\,000$, $v = 270$ and $F = 350\,000$, find the work done in a 10-second period by those forces which are doing work, and show that the work–energy principle holds in this case.

At a later time the pilot increases the thrust of the aircraft's engines to 400 000 N. When the aircraft has travelled a distance of 30 km, its speed has increased to $300\,\text{ms}^{-1}$.

(e) Find the work done against air resistance during this period, and the average resistance force.

Solution **(a)**

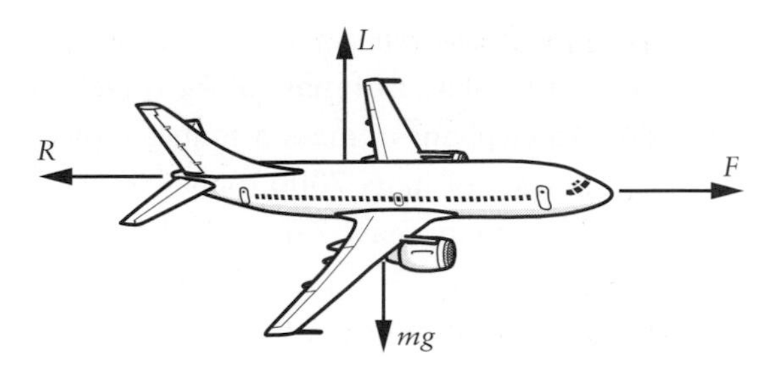

FIGURE 4.5

(b) Since the aircraft is travelling at constant velocity it is in equilibrium.

Horizontal forces $F = R$

Vertical forces $L = mg$

(c) Since the aircraft's velocity has no vertical component, the vertical forces, L and mg, are doing no work.

(d) In 10 s at $270\,\text{ms}^{-1}$ the aircraft travels 2700 m.

$$\text{Work done by force } F = 350\,000 \times 2700 = 94\,500\,000\,\text{J}$$
$$\text{Work done by force } R = 350\,000 \times -2700 = -94\,500\,000\,\text{J}$$

The work–energy principle states that in this situation

work done by F + work done by R = change in kinetic energy.

Now work done by F + work done by R = $(94\,500\,000 - 94\,500\,000) = 0\,\text{J}$, and change in kinetic energy = 0 (since velocity is constant), so the work–energy principle does indeed hold in this case.

(e)

$$\begin{aligned}
\text{Final K.E.} - \text{initial K.E.} &= \tfrac{1}{2}mv^2 - \tfrac{1}{2}mu^2 \\
&= \tfrac{1}{2} \times 100\,000 \times 300^2 - \tfrac{1}{2} \times 100\,000 \times 270^2 \\
&= 855 \times 10^6\,\text{J} \\
\text{Work done by driving force} &= 400\,000 \times 30\,000 \\
&= 12\,000 \times 10^6\,\text{J} \\
\text{Total work done} &= \text{K.E. gained} \\
\text{Work done by resistance force} + 12\,000 \times 10^6 &= 855 \times 10^6 \\
\text{Work done by resistance force} &= -11\,145 \times 10^6\,\text{J} \\
\text{Average force} \times \text{distance} &= \text{work done by force} \\
\text{Average force} \times 30\,000 &= -11\,145 \times 10^6
\end{aligned}$$

$\Rightarrow$ The average resistance force is 371 500 N (in the negative direction).

Note When an aircraft is in flight, most of the work done by the resistance force results in air currents and the generation of heat. A typical large jet cruising at 35 000 feet has a body temperature about 30°C above the surrounding air temperature. For supersonic flight the temperature difference is much greater. Supersonic aircraft fly with a skin temperature more than 200°C above that of the surrounding air.

EXERCISE 4A

1 Find the kinetic energy of the following objects.

(a) An ice skater of mass 50 kg travelling with speed 10 ms^{-1}.

(b) An elephant of mass 5 tonnes moving at 4 ms^{-1}.

(c) A train of mass 7000 tonnes travelling at 40 ms^{-1}.

(d) The moon, mass 7.4×10^{22} kg, travelling at 1000 ms^{-1} in its orbit round the earth.

(e) A bacterium of mass 2×10^{-16} g which has speed 1 mm s^{-1}.

2 Find the work done by a man in the following situations.

(a) He pushes a packing case of mass 35 kg a distance of 5 m across a rough floor against a resistance of 200 N. The case starts and finishes at rest.

(b) He pushes a packing case of mass 35 kg a distance of 5 m across a rough floor against a resistance force of 200 N. The case starts at rest and finishes with a speed of 2 ms^{-1}.

(c) He pushes a packing case of mass 35 kg a distance of 5 m across a rough floor against a resistance force of 200 N. Initially the case has speed 2 ms^{-1} but it ends at rest.

(d) He is handed a packing case of mass 35 kg. He holds it stationary, at the same height, for 20 s and then someone else takes it from him.

3 A sprinter of mass 60 kg is at rest at the beginning of a race and accelerates to 12 ms^{-1} in a distance of 30 m. Assume air resistance to be negligible.

(a) Calculate the kinetic energy of the sprinter at the end of the 30 m.

(b) Write down the work done by the sprinter over this distance.

(c) Calculate the forward force exerted by the sprinter, assuming it to be constant, using work = force $\times$ distance.

(d) Using force = mass $\times$ acceleration and the constant acceleration formulae, show that this force is consistent with the sprinter having speed 12 ms^{-1} after 30 m.

4 A sports car of mass 1.2 tonnes accelerates from rest to 30 ms^{-1} in a distance of 150 m. Assume air resistance to be negligible.

(a) Calculate the work done in accelerating the car. Does your answer depend on an assumption that the driving force is constant?

(b) If the driving force is in fact constant, what is its magnitude?

5 A car of mass 1600 kg is travelling at speed 25 ms^{-1} when the brakes are applied so that it stops after moving a further 75 m.

(a) Find the work done by the brakes.

(b) Find the retarding force from the brakes, assuming that it is constant and that other resistive forces may be neglected.

6 The forces acting on a hot air balloon of mass 500 kg are its weight and the total uplift force.

(a) Find the total work done when the speed of the balloon changes from

(i) $2\,\text{ms}^{-1}$ to $5\,\text{ms}^{-1}$ **(ii)** $8\,\text{ms}^{-1}$ to $3\,\text{ms}^{-1}$.

(b) If the balloon rises 100 m vertically while its speed changes calculate in each case the work done by the uplift force.

7 A bullet of mass 20 g, found at the scene of a police investigation, had penetrated 16 cm into a wooden post. The speed for that type of bullet is known to be $80\,\text{ms}^{-1}$.

(a) Find the kinetic energy of the bullet before it entered the post.

(b) What happened to this energy when the bullet entered the wooden post?

(c) Write down the work done in stopping the bullet.

(d) Calculate the resistive force on the bullet, assuming it to be constant.

Another bullet of the same mass and shape had clearly been fired from a different and unknown type of gun. This bullet had penetrated 20 cm into the post.

(e) Estimate the speed of this bullet before it hit the post.

8 The Highway Code gives the braking distance for a car travelling at $22\,\text{ms}^{-1}$ (50 mph) to be 38 m (125 ft). A car of mass 1300 kg is brought to rest in just this distance. It may be assumed that the only resistance forces come from the car's brakes.

(a) Find the work done by the brakes.

(b) Find the average force exerted by the brakes.

(c) What happened to the kinetic energy of the car?

(d) What happens when you drive a car with the handbrake on?

9 A car of mass 1200 kg experiences a constant resistance force of 600 N. The driving force from the engine depends upon the gear, as shown in the table.

Gear	1	2	3	4
Force (N)	2800	2100	1400	1000

Starting from rest, the car is driven 20 m in first gear, 40 m in second, 80 m in third and 100 m in fourth. How fast is the car travelling at the end?

10 In this question take g to be $10\,\text{ms}^{-2}$. A chest of mass 60 kg is resting on a rough horizontal floor. The coefficient of friction between the floor and the chest is 0.4. A woman pushes the chest in such a way that its speed–time graph is as shown below.

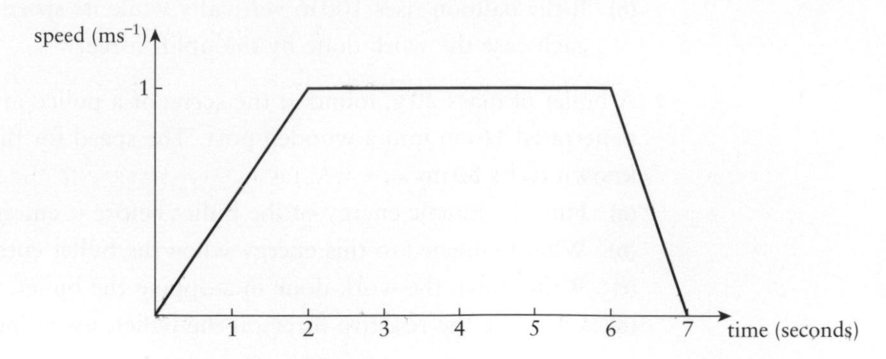

(a) Find the force of frictional resistance acting on the chest when it moves.

(b) Use the speed–time graph to find the total distance travelled by the chest.

(c) Find the total work done by the woman.

(d) Find the acceleration of the chest in the first 2 s of its motion and hence the force exerted by the woman during this time, and the work done.

(e) In the same way find the work done by the woman during the time intervals 2 to 6 s, and 6 to 7 s.

(f) Show that your answers to parts **(d)** and **(e)** are consistent with your answer to part **(c)**.

Gravitational potential energy

As you have seen, kinetic energy (K.E.) is the energy that an object has because of its motion. Potential energy (P.E.) is the energy an object has because of its position. The units of potential energy are the same as those of kinetic energy or any other form of energy, namely joules.

One form of potential energy is *gravitational potential energy*. The gravitational potential energy of the object in figure 4.6 of mass m kg at height h m above a fixed reference level, O, is mgh J. If it falls to the reference level, the force of gravity does mgh J of work and the body loses mgh J of potential energy.

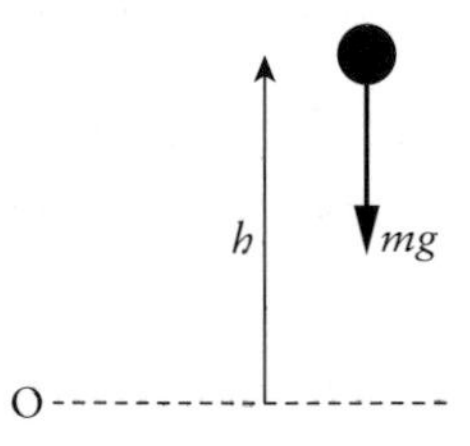

Figure 4.6

A loss in gravitational potential energy is an alternative way of accounting for the work done by the force of gravity.

If a mass m kg is *raised* through a distance h m, the gravitational potential energy *increases* by mgh J. If a mass m kg is lowered through a distance h m the gravitational potential energy *decreases* by mgh J.

EXAMPLE 4.8

Calculate the gravitational potential energy, relative to the ground, of a ball of mass 0.15 kg at a height of 2 m above the ground.

Solution Mass $m = 0.15$, height $h = 2$.

$$\begin{aligned}\text{Gravitational potential energy} &= mgh \\ &= 0.15 \times 9.8 \times 2 \\ &= 2.94\ \text{J}\end{aligned}$$

Note

If the ball falls:

$$\begin{aligned}\text{loss in P.E.} &= \text{work done by gravity} \\ &= \text{gain in K.E.}\end{aligned}$$

There is no change in the total energy (P.E. + K.E.) of the ball.

USING CONSERVATION OF MECHANICAL ENERGY

When gravity is the only force which does work on a body, mechanical energy is conserved. When this is the case, many problems are easily solved using energy. This is possible even when the acceleration is not constant.

EXAMPLE 4.9

A skier slides down a smooth ski slope 400 m long which is at an angle of 30° to the horizontal. Find the speed of the skier when he reaches the bottom of the slope.

At the foot of the slope the ground becomes horizontal and is made rough in order to help him to stop. The coefficient of friction between his skis and the ground is $\frac{1}{4}$.

(a) Find how far the skier travels before coming to rest.
(b) In what way is your model unrealistic?

Solution The skier is modelled as a particle.

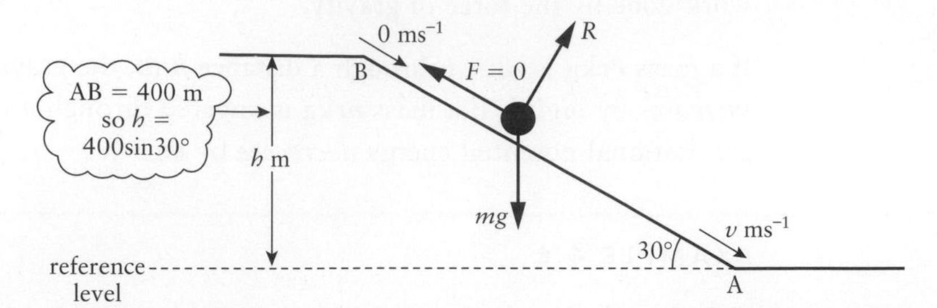

FIGURE 4.7

(a) Since in this case the slope is smooth, the frictional force is zero. The skier is subject to two external forces, his weight mg and the normal reaction from the slope.

The normal reaction between the skier and the slope does no work because the skier does not move in the direction of this force. The only force which does work is gravity, so mechanical energy is conserved.

$$\begin{aligned}\text{Total mechanical energy at B} &= mgh + \tfrac{1}{2}mu^2 \\ &= m \times 9.8 \times 400\sin 30^\circ + 0 \\ &= 1960m\,\text{J}\end{aligned}$$

$$\text{Total mechanical energy at A} = (0 + \tfrac{1}{2}mv^2)\,\text{J}$$

Since mechanical energy is conserved

$$\begin{aligned}\tfrac{1}{2}mv^2 &= 1960m \qquad ① \\ v^2 &= 3920 \\ v &= 62.6\end{aligned}$$

The skier's speed at the bottom of the slope is $62.6\,\text{ms}^{-1}$.

Notice that the mass of the skier cancels out. Using this model, all skiers should arrive at the bottom of the slope with the same speed. Also the slope could be curved so long as the total height lost is the same.

For the horizontal part there is some friction. Suppose that the skier travels a further distance s m before stopping.

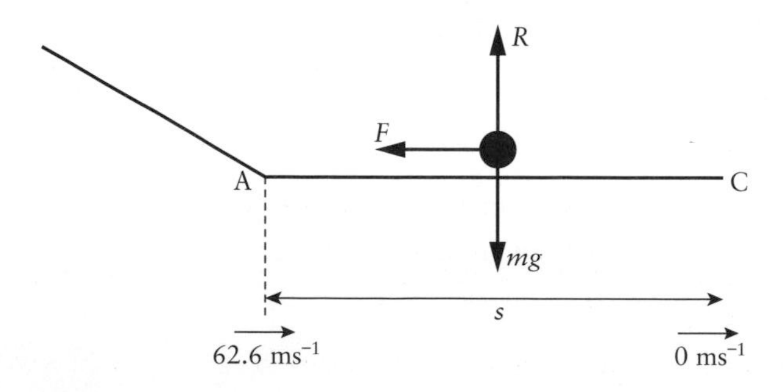

FIGURE 4.8

Coulomb's law of friction gives $\quad F = \mu R = \tfrac{1}{4}R.$

Since there is no vertical acceleration we can also say $R = mg$

So $$F = \tfrac{1}{4}mg$$

Work done by the friction force $F \times (-s) = -\tfrac{1}{4}mgs$.

Negative because the motion is in the opposite direction to the force

The increase in kinetic energy between A and C = $(0 - \tfrac{1}{2}mv^2)$ J.

Using the work–energy principle

$$-\tfrac{1}{4}mgs = -\tfrac{1}{2}mv^2 = -1960m \text{ from } ①$$

Solving for s gives $s = 800$.

So the distance the skier travels before stopping is 800 m.

(b) The assumptions made in solving this problem are that friction on the slope and air resistance are negligible, and that the slope ends in a smooth curve at A. Clearly the speed of 62.6 ms^{-1} is very high, so the assumption that friction and air resistance are negligible must be suspect.

EXAMPLE 4.10

Ama, whose mass is 40 kg, is taking part in an assault course. The obstacle shown in figure 4.9 is a river at the bottom of a ravine 8 m wide which she has to cross by swinging on a rope 5 m long secured to a point on the branch of a tree, immediately above the centre of the ravine.

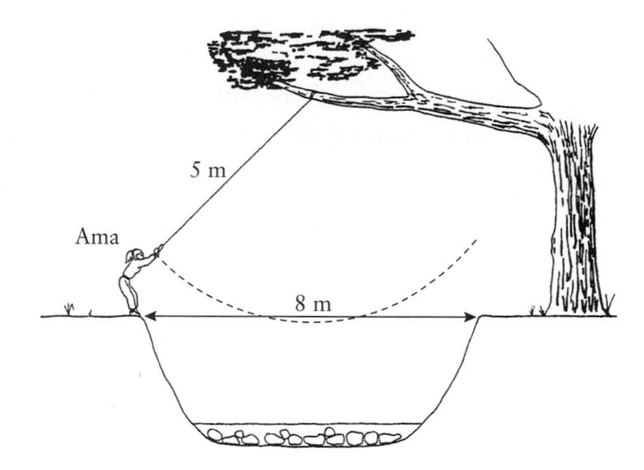

FIGURE 4.9

(a) If she starts from rest, find how fast Ama is travelling at the lowest point of her crossing.

(b) If she launches herself off at a speed of 1 ms^{-1}, will her speed be 1 ms^{-1} faster throughout her crossing?

Solution **(a)** The vertical height Ama loses is HB in the diagram.

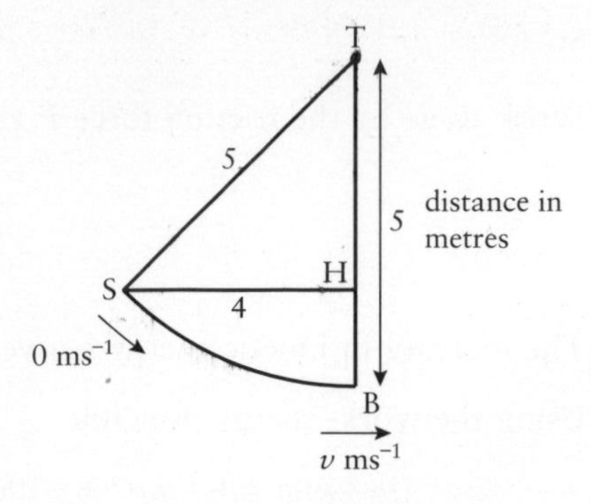

FIGURE 4.10

Using Pythagoras $\quad TH = \sqrt{5^2 - 4^2} = 3$

$$HB = 5 - 3 = 2$$

$$\text{P.E. lost} = mgh$$
$$= 40g \times 2$$
$$\text{K.E. gained} = \tfrac{1}{2}mv^2 - 0$$
$$= \tfrac{1}{2} \times 40 \times v^2$$

By conservation of energy, K.E. gained = P.E. lost

$$\tfrac{1}{2} \times 40 \times v^2 = 40 \times 9.8 \times 2$$
$$v = 6.26$$

Ama is travelling at $6.26\,\text{ms}^{-1}$.

(b) If she has initial speed $1\,\text{ms}^{-1}$ at S and speed $v\,\text{ms}^{-1}$ at B, her initial K.E. is $\tfrac{1}{2} \times 40 \times 1^2$ J and her K.E. at B is $\tfrac{1}{2} \times 40 \times v^2$.

Using conservation of energy

$$\tfrac{1}{2} \times 40 \times v^2 - \tfrac{1}{2} \times 40 \times 1^2 = 40 \times 9.8 \times 2$$

This gives $v = 6.34$, so Ama's speed at the lowest point is now $6.34\,\text{ms}^{-1}$, only $0.08\,\text{ms}^{-1}$ faster than in part **(a)**, so she clearly will not travel $1\,\text{ms}^{-1}$ faster throughout.

Historical note

James Joule was born in Salford in Lancashire on Christmas Eve 1818. He studied at Manchester University at the same time as the famous chemist, Dalton.

Joule spent much of his life conducting experiments to measure the equivalence of heat and mechanical forms of energy to ever increasing degrees of accuracy. Working with Thompson, he also discovered that a gas cools when it expands without doing work against external forces. It was this discovery that paved the way for the development of refrigerators.

Joule died in 1889 but his contribution to science is remembered with the S.I. Unit for energy named after him.

WORK AND KINETIC ENERGY FOR TWO-DIMENSIONAL MOTION

WORK DONE BY A FORCE AT AN ANGLE TO THE DIRECTION OF MOTION

You have probably noticed that as a cyclist you do work only against the component of the wind force that is directly against you. The sideways component does not resist your forward progress.

Suppose that you are sailing and the angle between the force, F, of the wind on your sail and the direction of your motion is θ. In a certain time you travel a distance d in the direction of F, see figure 4.11, but during that time you actually travel a distance s along the line OP.

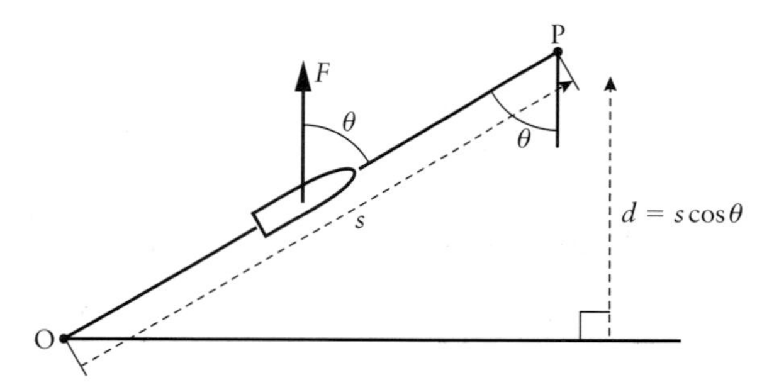

FIGURE 4.11

Work done by $F = Fd$

Since $d = s\cos\theta$, the work done by the force F is $Fs\cos\theta$. This can also be written as the product of the component of F along OP, $F\cos\theta$, and the distance moved along OP, s.

$$F \times s\cos\theta = F\cos\theta \times s$$

Notice that the direction of F is not necessarily the same as the direction of the wind; it depends on how you have set your sails.

EXAMPLE 4.11

As a car of mass m kg drives up a slope at an angle α to the horizontal it experiences a constant resistive force F N and a driving force D N. What can be deduced about the work done by D as the car moves a distance d m uphill if

(a) the car moves at constant speed?

(b) the car slows down?

(c) the car gains speed?

The initial and final speeds of the car are denoted by $u\,\text{ms}^{-1}$ and $v\,\text{ms}^{-1}$ respectively.

(d) Write v^2 in terms of the other variables.

Solution The diagram shows the forces acting on the car. The table shows the work done by each force. The normal reaction, R, does no work as the car moves no distance in the direction of R.

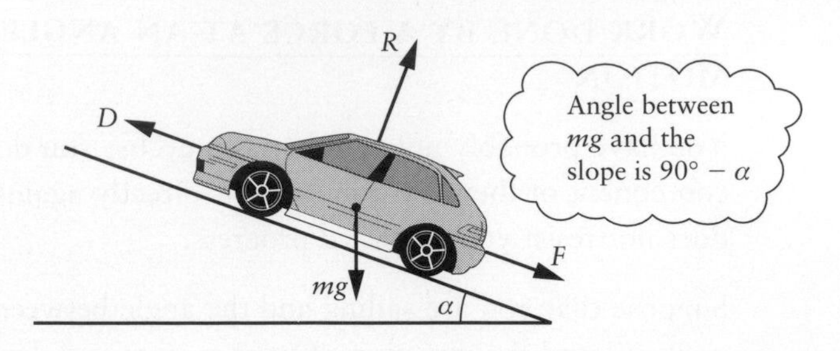

FIGURE 4.12

Force	Work done
Resistance F	$-Fd$
Normal reaction R	0
Force of gravity mg	$-mgd\cos(90° - \alpha) = -mgd\sin\alpha$
Driving force D	Dd
Total work done	$Dd - Fd - mgd\sin\alpha$

(a) If the car moves at a constant speed there is no change in kinetic energy so the total work done is zero, giving

Work done by D is

$$Dd = Fd + mgd\sin\alpha$$

(b) If the car slows down the total work done by the forces is negative, hence

Work done by D is

$$Dd < Fd + mgd\sin\alpha$$

(c) If the car gains speed the total work done by the forces is positive

Work done by D is

$$Dd > Fd + mgd\sin\alpha$$

(d) Total work done = final K.E. – initial K.E.

$$\Rightarrow \quad Dd - Fd - mgd\sin\alpha = \tfrac{1}{2}mv^2 - \tfrac{1}{2}mu^2$$

Multiplying by $\dfrac{2}{m}$

$$\Rightarrow \quad v^2 = u^2 + \frac{2d}{m}(D - F) - 2gd\sin\alpha$$

EXERCISE 4B

1 Calculate the gravitational potential energy, relative to the reference level OA, for each of the objects shown.

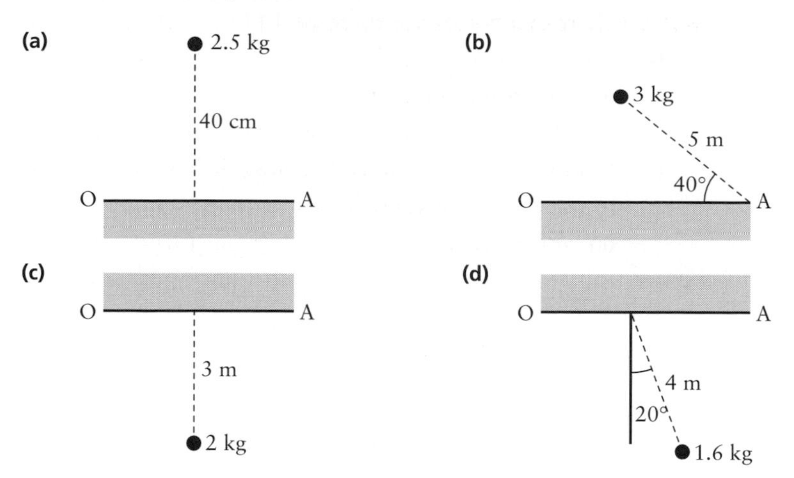

2 Calculate the change in gravitational potential energy when each object moves from A to B in the situations shown below. State whether the change is an increase or a decrease.

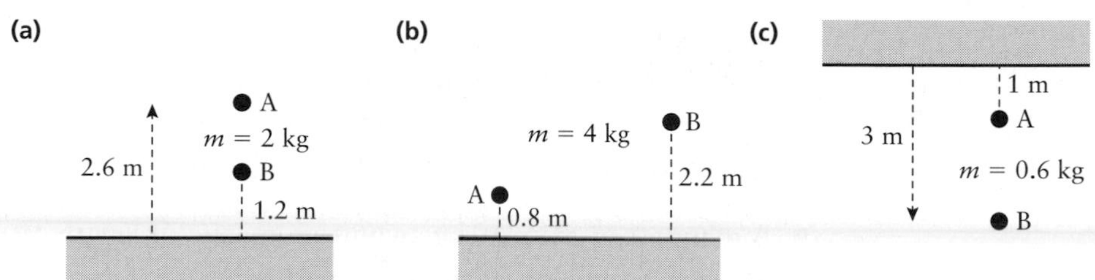

3 A vase of mass 1.2 kg is lifted from ground level and placed on a shelf at a height of 1.5 m. Find the work done against the force of gravity.

4 Find the increase in gravitational potential energy of a woman of mass 60 kg who climbs to the twelfth floor of a block of flats. The distance between floors is 3.3 m.

5 A car of mass 0.9 tonnes is driven 200 m up a slope inclined at 5° to the horizontal. There is a resistance force of 100 N.

(a) Find the work done by the car against gravity.

(b) Find the work done against the resistance force.

(c) When asked to work out the total work done by the car, a student replied '$(900g + 100) \times 200$ J'. Explain the error in this answer.

(d) If the car slows down from $12\,\text{ms}^{-1}$ to $8\,\text{ms}^{-1}$, what is the total work done by the engine?

6 A sledge of mass 10 kg is being pulled across level ground by a rope which makes an angle of 20° with the horizontal. The tension in the rope is 80 N and there is a resistance force of 14 N.

(a) Find the work done by

(i) the tension in the rope

(ii) the resistance force while the sledge moves a distance of 20 m.

(b) Find the speed of the sledge after it has moved 20 m

(i) if it starts at rest

(ii) if it starts at $4\,\text{ms}^{-1}$.

7 A bricklayer carries a hod of bricks of mass 25 kg up a ladder of length 10 m inclined at an angle of 60° to the horizontal.

(a) Calculate the increase in the gravitational potential energy of the bricks.

(b) If instead he had raised the bricks vertically to the same height, using a rope and pulleys, would the increase in potential energy be **(i)** less, **(ii)** the same, or **(iii)** more than in part **(a)**?

8 A girl of mass 45 kg slides down a smooth water chute of length 6 m inclined at an angle of 40° to the horizontal. She starts from rest.

(a) Find

(i) the decrease in her potential energy

(ii) her speed at the bottom.

(b) How are answers to part **(a)** affected if the slide is not smooth?

9 A gymnast of mass 50 kg swings on a rope of length 10 m. Initially the rope makes an angle of 50° with the vertical.

(a) Find the decrease in her potential energy when the rope has reached the vertical.

(b) Find her kinetic energy and hence her speed when the rope is vertical, assuming that air resistance may be neglected.

(c) The gymnast continues to swing. What angle will the rope make with the vertical when she is next temporarily at rest?

(d) Explain why the tension in the rope does no work.

10 A stone of mass 0.2 kg is dropped from the top of a building 78.4 m high. After t s it has fallen a distance x m and has speed $v\,\text{ms}^{-1}$.

(a) What is the gravitational potential energy of the stone relative to ground level when it is at the top of the building?

(b) What is the potential energy of the stone t s later?

(c) Show that, for certain values of t, $v^2 = 19.6x$ and state the range of values of t for which it is true.

(d) Find the speed of the stone when it is half-way to the ground.

(e) At what height will the stone have half its final speed?

11 Wesley, whose mass is 70 kg, inadvertently steps off a bridge 50 m above water. When he hits the water, Wesley is travelling at 25 ms^{-1}.

(a) Calculate the potential energy Wesley has lost and the kinetic energy he has gained.

(b) Find the size of the resistance force acting on Wesley while he is in the air, assuming it to be constant.

Wesley descends to a depth of 5 m below the water surface, then returns to the surface.

(c) Find the total upthrust (assumed constant) acting on him while he is moving downwards in the water.

12 A hockey ball of mass 0.15 kg is hit from the centre of a pitch. Its position vector (in m) t s later is modelled by

$$\mathbf{r} = 10t\mathbf{i} + (10t - 4.9t^2)\mathbf{j}$$

where the unit vectors **i** and **j** are in directions along the line of the pitch and vertically upwards.

(a) What value of g is used in this model?

(b) Find an expression for the gravitational potential energy of the ball at time t. For what values of t is your answer valid?

(c) What is the maximum height of the ball? What is its velocity at that instant?

(d) Find the initial velocity, speed and kinetic energy of the ball.

(e) Show that according to this model mechanical energy is conserved and state what modelling assumption is implied by this. Is it reasonable in this context?

13 A ski-run starts at altitude 2471 m and ends at 1863 m.

(a) If all resistance forces could be ignored, what would the speed of the skier be at the end of the run?

A particular skier of mass 70 kg actually attains a speed of 42 ms^{-1}. The length of the run is 3.1 km.

(b) Find the average force of resistance acting on a skier.

Two skiers are equally skilful.

(c) Which would you expect to be travelling faster by the end of the run, the heavier or the lighter?

14 A tennis ball of mass 0.06 kg is hit vertically upwards with speed 20 ms^{-1} from a point 1.1 m above the ground. It reaches a height of 16 m.

(a) Find the initial kinetic energy of the ball, and its gain in potential energy when it is at its highest point

(b) Calculate the loss of mechanical energy due to air resistance.

(c) Find the magnitude of the air resistance force on the ball, assuming it to be constant while the ball is moving.

(d) With what speed does the ball land?

15 Akosua draws water from a well 12 m below the ground. Her bucket holds 5 kg of water and by the time she has pulled it to the top of the well it is travelling at $1.2\,\text{ms}^{-1}$.

(a) How much work does Akosua do in drawing the bucket of water?

On an average day 150 people in the village each draw six such buckets of water. One day a new electric pump is installed that takes water from the well and fills an overhead tank 5 m above ground level every morning. The flow rate through the pump is such that the water has speed $2\,\text{ms}^{-1}$ on arriving in the tank.

(b) Assuming that the villagers' demand for water remains unaltered, how much work does the pump do in 1 day?

It takes the pump 1 hour to fill the tank each morning.

(c) At what rate does the pump do work, in joules per second (watts)?

16

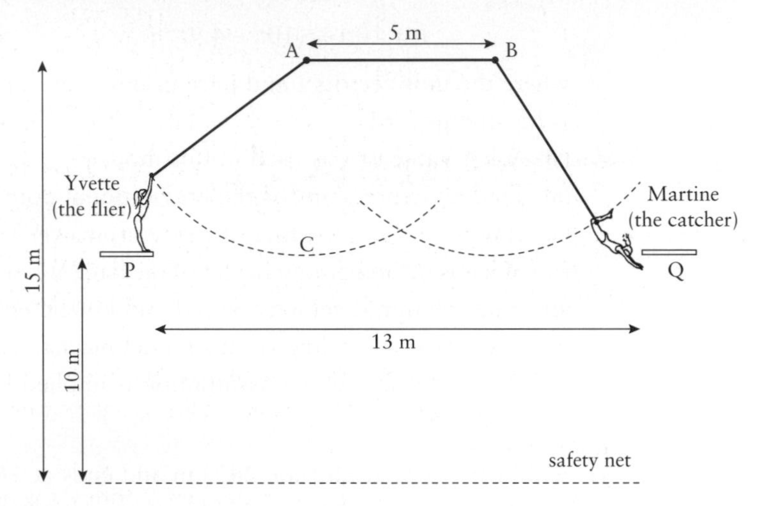

Yvette and Martine are trapeze artists in a circus. Their equipment is shown in the diagram. The trapezes are suspended from points A and B, 5 m apart but at the same height, 15 m above the safety net. The two platforms P and Q are at the same height, 10 m above the safety net, 13 m apart and placed symmetrically with respect to points A and B. C is level with P and Q.

Yvette, who is the 'flier', holds her trapeze while standing on the edge of platform P, with her arms straight above her. In this position her hands are 2 m above the platform.

(a) How long is the rope of Yvette's trapeze?

(b) How fast is Yvette travelling when she passes the lowest point (C) of her trapeze's arc? (Assume she does not push herself off.)

As part of their act they frighten the audience. Yvette lets go of her trapeze at point C and Martine pretends to forget to catch her. Yvette falls into the safety net.

(c) How far will Yvette move horizontally before her feet land in the net?

(d) Describe, and comment on, the assumptions you have made in modelling this situation.

POWER

It is claimed that a motorcycle engine can develop a maximum *power* of 26.5 kW at a top *speed* of 103 mph. This suggests that power is related to speed and this is indeed the case.

Power is the rate at which work is being done. A powerful car does work at a greater rate than a less powerful one.

You might find it helpful to think in terms of a force, F, acting for a very short time t over a small distance s. Assume F to be constant over this short time. Power is the rate of working so

$$\text{power} = \frac{\text{work}}{\text{time}} = \frac{Fs}{t} = Fv$$

This gives you the power at an *instant* of time. The result is true whether or not F is constant

The power of a vehicle moving at speed v under a driving force F is given by Fv.

For a motor vehicle the power is produced by the engine, whereas for a bicycle it is produced by the cyclist. They both make the wheels turn, and the friction between the rotating wheels and the ground produces a forward force on the machine.

The unit of power is the watt (W), named after James Watt. The power produced by a force of 1 N acting on an object that is moving at 1 ms^{-1} is 1 W. Because the watt is such a small unit you will probably use kilowatts more often (1 kW = 1000 W).

EXAMPLE 4.12

A car of mass 1000 kg can produce a maximum power of 45 kW. Its driver wishes to overtake another vehicle. Ignoring air resistance, find the maximum acceleration of the car when it is travelling at

(a) 12 ms^{-1} **(b)** 28 ms^{-1} (these are about 27 mph and 63 mph).

Solution

(a) Power = force × velocity
The driving force at 12 ms^{-1} is F_1 N where

$$45\,000 = F_1 \times 12$$
$$\Rightarrow \quad F_1 = 3750$$

By Newton's second law $F = ma$

$$\Rightarrow \quad \text{acceleration} = \frac{3750}{1000} = 3.75\ \text{ms}^{-2}$$

(b) Now the driving force F_2 is given by

$$45\,000 = F_2 \times 28$$
$$\Rightarrow \quad F_2 = 1607$$
$$\Rightarrow \quad \text{acceleration} = \frac{1607}{1000} = 1.61\ \text{ms}^{-2}$$

This example shows why it is easier to overtake a slow-moving vehicle.

EXAMPLE 4.13

A car of mass 900 kg produces power 45 kW when moving at a constant speed. It experiences a resistance of 1700 N.

(a) What is its speed?
(b) The car comes to a downhill stretch inclined at 2° to the horizontal. What is its maximum speed downhill if the power and resistance remain unchanged?

Solution **(a)** As the car is travelling at a constant speed, there is no resultant force on the car. In this case the forward force of the engine must have the same magnitude as the resistance forces, i.e. 1700 N.

Denoting the speed of the car by $v\,\text{ms}^{-1}$, $P = Fv$ gives

$$\begin{aligned} v &= \frac{P}{F} \\ &= \frac{45\,000}{1700} \\ &= 26.5 \end{aligned}$$

The speed of the car is $26.5\,\text{ms}^{-1}$ (approximately 60 mph).

(b) The diagram shows the forces acting.

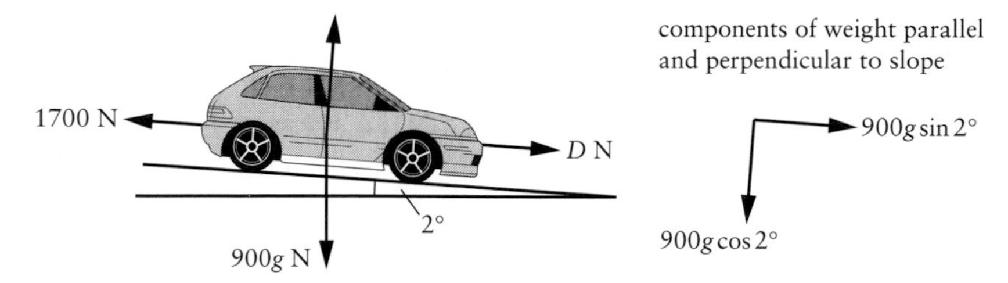

FIGURE 4.13

At maximum speed there is no acceleration so the resultant force down the slope is zero.

When the driving force is D N

$$D + 900g\sin 2° - 1700 = 0$$

$\Rightarrow$ $$D = 1392$$

But power is Dv so $$45\,000 = 1392v$$

$$\Rightarrow \quad v = \frac{45\,000}{1392}$$

The maximum speed is $32.3\,\text{ms}^{-1}$ (about 73 mph).

Historical note

James Watt was born in 1736 in Greenock in Scotland, the son of a house- and ship-builder. As a boy James was frail and he was taught by his mother rather than going to school. This allowed him to spend time in his father's workshop where he developed practical and inventive skills.

As a young man he manufactured mathematical instruments: quadrants, scales, compasses and so on. One day he was repairing a model steam engine for a friend and noticed that its design was very wasteful of steam. He proposed an alternative arrangement, which was to become standard on later steam engines. This was the first of many engineering inventions which made possible the subsequent industrial revolution. James Watt died in 1819, a well-known and highly respected man. His name lives on as the S.I. unit for power.

EXERCISE 4C

1 A builder hoists bricks up to the top of the house he is building. Each brick weighs 3.5 kg and the house is 9 m high. In the course of 1 hour the builder raises 120 bricks from ground level to the top of the house, where they are unloaded by his mate.

(a) Find the increase in gravitational potential energy of one brick when it is raised in this way.

(b) Find the total work done by the builder in 1 hour of raising bricks.

(c) Find the average power with which he is working.

2 A weightlifter takes 2 seconds to lift 120 kg from the floor to a position 2 m above it where the weight has to be held stationary.

(a) Calculate the work done by the weightlifter.

(b) Calculate the average power developed by the weightlifter.

The weightlifter is using the 'clean and jerk' technique. This means that in the first stage of the lift he raises the weight 0.8 m from the floor in 0.5 s. He then holds it stationary for 1 s before lifting it up to the final position in another 0.5 s.

(c) Find the average power developed by the weightlifter during each of the stages of the lift.

3 A winch is used to pull a crate of mass 180 kg up a rough slope of angle 30° against a frictional force of 450 N. The crate moves at a steady speed, v, of $1.2\ \text{ms}^{-1}$.

(a) Calculate the gravitational potential energy given to the crate during 30 s.

(b) Calculate the work done against friction during this time.

(c) Calculate the total work done per second by the winch.

The cable from the winch to the crate runs parallel to the slope.

(d) Calculate the tension, T, in the cable.

(e) What information is given by $T \times v$?

4 The power output from the engine of a car of mass 800 kg which is travelling along level ground at a constant speed of $33\ \text{ms}^{-1}$ is 23 200 W.

(a) Find the total resistance on the car under these conditions.

(b) You were given one piece of unnecessary information. Which is it?

5 A Kawasaki GPz 305 motorcycle has a maximum power output of 26.5 kW and a top speed of 103 mph ($46\ \text{ms}^{-1}$). Find the force exerted by the motorcycle engine when the motorcycle is travelling at top speed.

6 A crane is raising a load of 500 tonnes at a steady rate of $5\ \text{cm s}^{-1}$. What power is the engine of the crane producing? (Assume that there are no forces from friction or air resistance.)

7 A cyclist, travelling at a constant speed of $8\ \text{ms}^{-1}$ along a level road, experiences a total resistance of 70 N.

(a) Find the power which the cyclist is producing.

(b) Find the work done by the cyclist in 5 minutes under these conditions.

8 A conveyor belt picks up stationary sacks of grain and delivers them to a place 5 m higher with speed $1.5\,\text{ms}^{-1}$. The mass of one sack is 25 kg and they are delivered at the rate of one sack every 6 s.

(a) Calculate the total mechanical energy given to one sack by the conveyor belt.

(b) Calculate the average power with which the conveyor's belt is working, assuming that frictional forces may be ignored.

9 A train consists of a diesel shunter of mass 100 tonnes pulling a truck of mass 25 tonnes along a level track. The engine is working at a rate of 125 kW. The resistance to motion of the truck and shunter is 50 N per tonne.

(a) Calculate the constant speed of the train.

While travelling at this constant speed, the truck becomes uncoupled. The shunter engine continues to produce the same power.

(b) Find the acceleration of the shunter immediately after this happens.

(c) Find the greatest speed the shunter can now reach.

10 A supertanker of mass 4×10^8 kg is steaming at a constant speed of $8\,\text{ms}^{-1}$. The resistance force is 2×10^6 N.

(a) What power are the ship's engines producing?

One of the ship's two engines suddenly fails but the other continues to work at the same rate.

(b) Find the deceleration of the ship immediately after the failure.

The resistance force is directly proportional to the speed of the ship.

(c) Find the eventual steady speed of the ship under one engine only, assuming that the single engine maintains constant power output.

11 A car of mass 850 kg has a maximum speed of $50\,\text{ms}^{-1}$ and a maximum power output of 40 kW. The resistance force, R N, at speed $v\,\text{ms}^{-1}$ is modelled by

$$R = kv.$$

(a) Find the value of k.

(b) Find the resistance force when the car's speed is $20\,\text{ms}^{-1}$.

(c) Find the power needed to travel at a constant speed of $20\,\text{ms}^{-1}$ along a level road.

(d) Find the maximum acceleration of the car when it is travelling at $20\,\text{ms}^{-1}$

(i) along a level road

(ii) up a hill at $5°$ to the horizontal.

EXERCISE 4D **Examination-style questions**

Questions about collisions often include considerations of energy. Examples of these will be found in Chapter 5.

1 A particle of mass 800 g slides on a rough horizontal plane. It slows from 10 ms^{-1} to 5 ms^{-1} over a distance of 50 m. Find the friction force, assumed constant.

2 A particle is launched at 3 ms^{-1} up a line of greatest slope of a smooth plane inclined at 20° to the horizontal. Find the distance that it moves up the plane before coming to instantaneous rest.

3 An aircraft of mass 50 tonnes is climbing at an angle of 10° to the horizontal. Its speed is 90 ms^{-1} and the resistances to motion total 40 000 N. Find, in kW, the rate of working of the aircraft's engines.

4 A car of mass 800 kg is driving from A to B along an undulating road. The distance AB is 1500 m and B is 60 m higher than A. The driving force produced by the engine is 900 N and the resistances to motion total 650 N. The car passes A at a speed of 20 ms^{-1}. Find its speed at B.

5 A conveyor belt is moving sand from rest. It moves 10 kg of sand every second to a height of 3 m and delivers it with speed 2 ms^{-1}.

(a) Find the power needed to run the conveyor belt.

(b) What modelling assumptions have you made in your answer to part **(a)**?

6 A fountain is designed to float on a pond and is powered by solar cells which produce a power of 0.6 W. The water is pumped through a circular nozzle of diameter 1 cm and projected vertically. Assuming that there are no losses,

(a) find the speed with which the water leaves the nozzle

(b) find the height to which the water rises.

[1 m^3 of water has mass 1000 kg.]

7 A particle of mass 3 kg slides from rest down a rough plane inclined at an angle α to the horizontal, where $\sin\alpha = \frac{1}{7}$. The friction force between the particle and the plane has the value 4 N.

(a) Find the velocity of the particle after it has travelled 30 m.

At this point the angle that the plane makes with the horizontal changes to β, where $\sin\beta = \frac{1}{14}$. The friction is unchanged.

(b) Find how much further the particle slides before coming to rest.

8 A hang glider of mass 110 kg (including pilot) is descending in a straight line at an angle of 1° to the horizontal. Over a distance of 2 km along its flight, its speed increases from 10 ms^{-1} to 12 ms^{-1}.

(a) Find the change in the *total* energy over this distance, stating whether it is an increase or a decrease.

(b) Suggest a reason for the change.

9 A 60 kg pupil is running up a flight of stairs. The flight of stairs is 5 m long and rises a vertical height of 3 m, and the pupil takes 6 s to make the ascent.

(a) Ignoring resistances, find an estimate for the rate at which the pupil is working.

In fact there is a resistance force of R N, and the pupil is working at a rate P W.

(b) Express P in terms of R.

The pupil can descend at constant speed without doing any work.

(c) Find P and R.

10 A car of mass 1080 kg has an engine of power 60 kW. The resistances to motion are kv N, where v ms^{-1} is the speed of the car. The car's maximum speed on level ground is 50 ms^{-1}.

(a) Show that $k = 24$.

(b) Find the maximum acceleration of the car when it is travelling at a speed of 40 ms^{-1}.

(c) Find the maximum speed at which the car can climb a hill inclined at an angle α to the horizontal, where $\sin\alpha = \frac{1}{7}$.

11 A lorry of mass 5000 kg moves at a constant speed of 15 ms^{-1} up a hill inclined at an angle α to the horizontal, where $\sin\alpha = \frac{1}{15}$. The resistance experienced by the lorry is constant and has magnitude 2500 N. Find, in kW, the rate of working of the lorry's engine.

[Edexcel]

12 A woman of mass 60 kg runs along a horizontal track at a constant speed of 4 ms^{-1}. In order to overcome air resistance, she works at a constant rate of 120 W.

(a) Find the magnitude of the air resistance which she experiences.

She now comes to a hill inclined at an angle α to the horizontal where $\sin\alpha = \frac{1}{15}$. To allow for the hill, she reduces her speed to 3 ms^{-1} and maintains this constant speed as she runs up the slope. In a preliminary model of this situation, the air resistance is modelled as having the constant value obtained in part **(a)** whatever the speed of the woman.

(b) Estimate the rate at which the woman has to work against the external forces in order to run up the hill.

In a more refined model, the air resistance experienced by the woman is taken as proportional to the square of her speed.

(c) Use your answer to part **(a)** to obtain a revised estimate of the air resistance experienced by the woman when running at 3 ms^{-1}.

(d) Find a revised estimate of the rate at which the woman has to work against the external forces as she runs up the hill.

[Edexcel]

13 A car of mass 1500 kg is towing a caravan of mass 500 kg, by means of a light inextensible towbar, up a straight road which is inclined at an angle α to the horizontal, where $\sin\alpha = \frac{1}{10}$. The towbar is parallel to the road. The car's engine is working at a constant rate of 70 kW. The resistances to motion due to non-gravitational forces are constant, having magnitude 1000 N for the car and 400 N for the caravan.

At a given instant, the car and caravan are both moving with a speed of $10\,\text{m}\,\text{s}^{-1}$. Find

(a) the magnitude of the acceleration of the caravan

(b) the tension in the towbar.

At this instant the towbar breaks. By considering the change in kinetic energy of the caravan, or otherwise,

(c) find, in m to 3 significant figures, the further distance moved by the caravan before it comes momentarily to rest.

[Edexcel]

KEY POINTS

1 The work done by a constant force F is given by Fs where s is the distance moved in the direction of the force.

2 The kinetic energy (K.E.) of a body of mass m moving with speed v is given by $\frac{1}{2}mv^2$. Kinetic energy is the energy a body possesses on account of its motion.

3 The work–energy principle states that the total work done by all the forces acting on a body is equal to the increase in the kinetic energy of the body.

4 The gravitational potential energy of a body of mass m at height h above a given reference level is given by mgh. It is the work done against the force of gravity in raising the body.

5 Mechanical energy (K.E. and P.E.) is conserved when no forces other than gravity do work.

6 Power is the rate of doing work, and is given by Fv.

7 The S.I. unit for energy is the joule and that for power is the watt.

Chapter five

COLLISIONS

I never think I have hit hard, unless it rebounds.

Samuel Johnson

MOMENTUM AS A VECTOR

In *Mechanics 1*, you learnt about linear momentum in one direction only. In the first part of this chapter, you will extend this work to two dimensions.

Both impulse and momentum are vectors. The impulse of a force is in the direction of the force and the momentum of a moving object is in the direction of its velocity. When an impulse $\mathbf{J}$ changes the velocity of a mass m from $\mathbf{u}$ to $\mathbf{v}$, the impulse–momentum equation is

$$\mathbf{J} = m\mathbf{v} - m\mathbf{u}$$

The diagram shows how this applies to a ball which changes direction when it is hit by a bat.

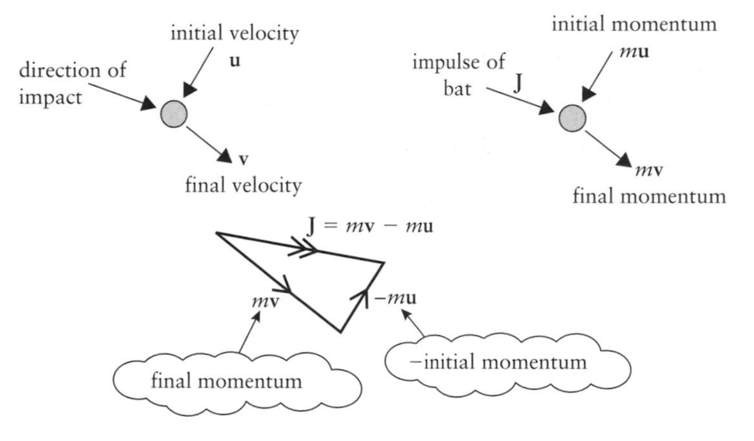

FIGURE 5.1

EXAMPLE 5.1

In a game of snooker the cue ball (W) of mass 0.2 kg is hit towards a stationary red ball (R) at $0.8\,\mathrm{m\,s^{-1}}$. After the collision the cue ball is moving at $0.6\,\mathrm{m\,s^{-1}}$ having been deflected through 30°.

FIGURE 5.2

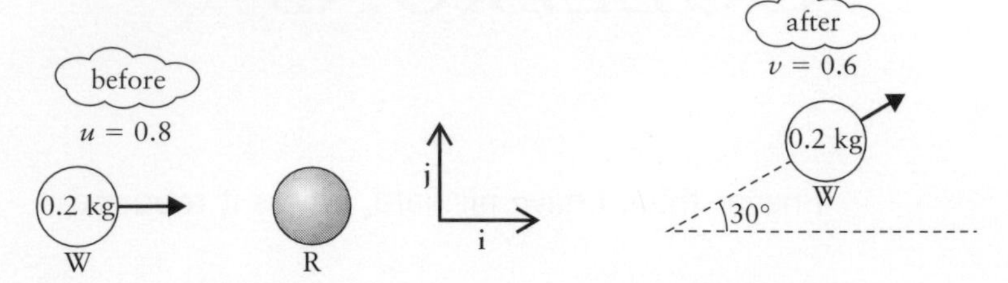

Find the impulse of the cue ball and show this in a vector diagram.

Solution In terms of unit vectors **i** and **j** the velocities before and after the collision are given by

$$\mathbf{u} = 0.8\mathbf{i}$$
$$\mathbf{v} = 0.6\cos 30°\,\mathbf{i} + 0.6\sin 30°\,\mathbf{j}$$

Then impulse = final momentum − initial momentum

$$\mathbf{J} = m\mathbf{v} - m\mathbf{u}$$
$$= 0.2(0.6\cos 30°\,\mathbf{i} + 0.6\sin 30°\,\mathbf{j}) - 0.2(0.8\mathbf{i})$$
$$= -0.056\,\mathbf{i} + 0.06\,\mathbf{j}$$

$$\text{Magnitude of impulse} = \sqrt{0.056^2 + 0.06^2}$$
$$= 0.082$$

$$\text{Direction: } \tan\alpha = \frac{0.06}{0.056}$$
$$\alpha = 47°$$
$$\theta = 133°$$

0.06, 0.056, α, θ

The impulse has magnitude 0.082 N at an angle of 133° to the initial motion of the ball.

This is shown on the vector diagram below. Note that the impulse–momentum equation shows the direction of the impulsive force acting on the cue ball.

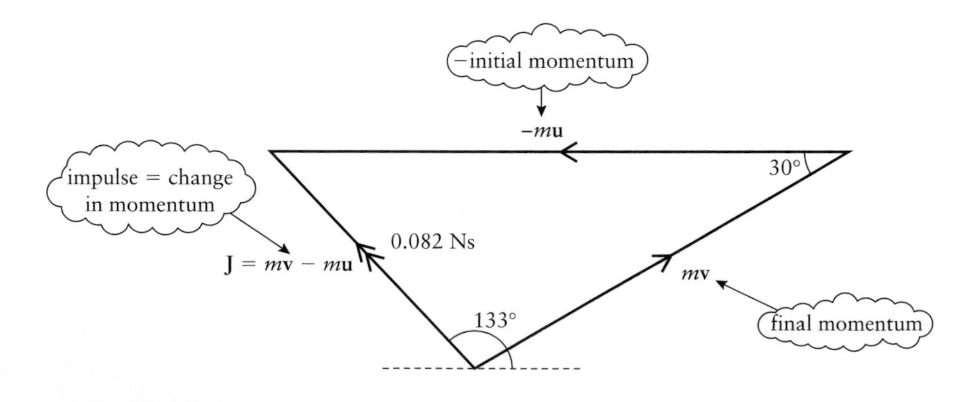

FIGURE 5.3

EXAMPLE 5.2

A hockey ball of mass 0.15 kg is moving at $4\,\text{ms}^{-1}$ parallel to the side of a pitch when it is struck by a blow from a hockey stick that exerts an impulse of 4 Ns at an angle of 120° to its direction of motion. Find the final velocity of the ball.

Solution The vector diagram shows the motion of the ball.

FIGURE 5.4

In terms of unit vectors **i** and **j** we have

$$\mathbf{u} = 4\mathbf{i}$$

and
$$\mathbf{J} = -4\cos 60°\mathbf{i} + 4\sin 60°\mathbf{j}$$
$$= -2\mathbf{i} + 3.46\mathbf{j}$$

Using
$$\mathbf{J} = m\mathbf{v} - m\mathbf{u}$$
$$-2\mathbf{i} + 3.46\mathbf{j} = 0.15\mathbf{v} - 0.15 \times 4\mathbf{i}$$
$$\Rightarrow \quad 0.15\mathbf{v} = -2\mathbf{i} + 0.6\mathbf{i} + 3.46\mathbf{j}$$
$$\Rightarrow \quad 0.15\mathbf{v} = -1.4\mathbf{i} + 3.46\mathbf{j}$$
$$\Rightarrow \quad \mathbf{v} = -9.33\mathbf{i} + 23.1\mathbf{j}$$

The magnitude of the velocity is given by $v = \sqrt{9.33^2 + 23.1^2}$
$= 24.9\,\text{ms}^{-1}$

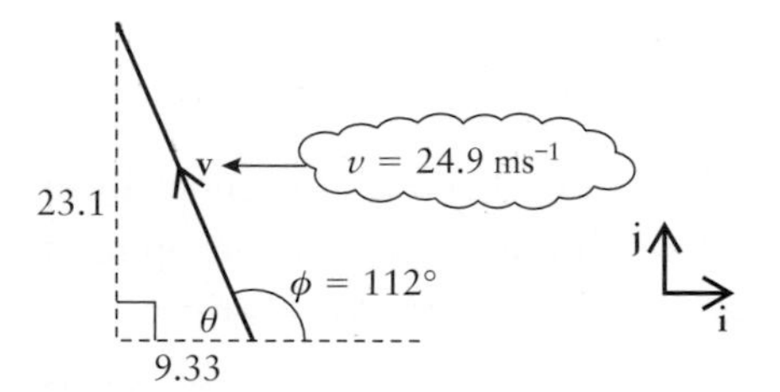

FIGURE 5.5

The angle, θ, is given by $\theta = \tan^{-1}\left(\frac{23.1}{9.33}\right)$

$$= 68°$$
$$\phi = 180° - 68°$$
$$= 112°$$

After the blow, the ball has a velocity of magnitude $24.9\,\text{ms}^{-1}$ at an angle of 112° to the original direction of motion.

EXERCISE 5A

1 A snooker ball of mass 0.08 kg is travelling with speed 3.5 ms^{-1} when it hits the cushion at an angle of 60°. After the impact the ball is travelling with speed 2 ms^{-1} at an angle 30° to the cushion.

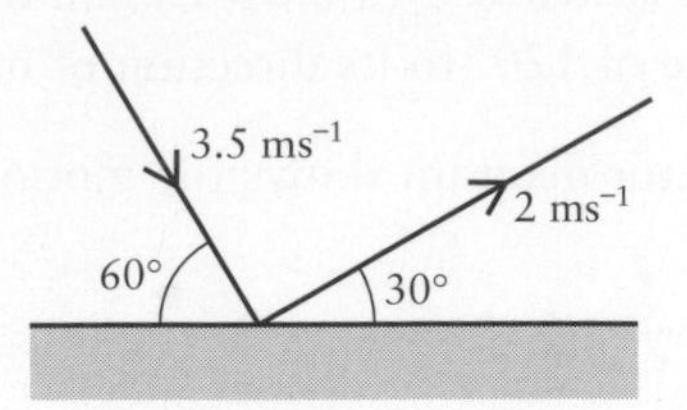

(a) Draw accurate scale diagrams to represent the following vectors:
   (i) the momentum of the ball before impact
   (ii) the momentum of the ball after impact
   (iii) the change in momentum of the ball during impact.

(b) Use your answer to part **(a) (iii)** to *estimate* the magnitude and direction of the impulse acting on the ball.

(c) Resolve the velocity of the ball before and after impact into components parallel and perpendicular to the cushion.

(d) Use your answers to part **(c)** to *calculate* the impulse which acts on the ball during its impact with the cushion. Comment on your answers.

2 A hockey ball of mass 0.15 kg is travelling with velocity $12\mathbf{i} - 8\mathbf{j}$ (in ms^{-1}), where the unit vectors **i** and **j** are in horizontal directions parallel and perpendicular to the length of the pitch. The ball is hit by Jane with an impulse $-4.8\mathbf{i} + 1.2\mathbf{j}$.

(a) What is the velocity of the ball immediately after Jane has hit it?

The ball goes straight, without losing any speed, to Fatima in the opposite team who hits it without stopping it. Its velocity is now $14\mathbf{i} + 7\mathbf{j}$.

(b) What impulse does Fatima give the ball?

(c) Which player hits the ball harder?

3 A hailstone of mass 4 g is travelling with speed 20 ms^{-1} when it hits a window as shown in the diagram. It bounces off the window; the vertical component of the velocity is unaltered, but the horizontal component is now 2 ms^{-1} away from the window.

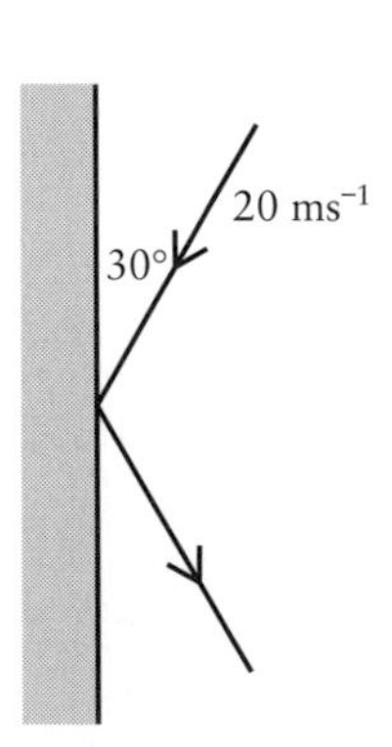

(a) State the magnitude and direction of the impulse
   (i) of the window on the hailstone
   (ii) of the hailstone on the window.

At the peak of the storm, hailstones like this are hitting the window at the rate of 540 per minute.

(b) Find the average force of the hail on the window.

NEWTON'S LAW OF IMPACT

If you drop two different balls, say a tennis ball and a cricket ball, from the same height, will they both rebound to the same height? How will the heights of the second bounces compare with the heights of the first ones?

Your own experience probably tells you that different balls will rebound to different heights. For example, a tennis ball will rebound to a greater height than a cricket ball. Furthermore, the surface on which the ball is dropped will affect the bounce. A tennis ball dropped on to a concrete floor will rebound higher than if dropped on to a carpeted floor. The following experiment allows you to look at this situation more closely.

EXPERIMENT

The aim of this experiment is to investigate what happens when balls bounce. Make out a table to record your results.

1. Drop a ball from a variety of heights and record the heights of release h_a and rebound h_s. Repeat several times for each height.
2. Use your values of h_a and h_s, to calculate v_a and v_s, the speeds on impact and rebound. Enter the results in your table.
3. Calculate the ratio $\frac{v_a}{v_s}$ for each pair of readings of h_a and h_s and enter the results in your table.
4. What do you notice about these ratios?
5. Repeat the experiment with different types of ball.

COEFFICIENT OF RESTITUTION

Newton's experiments on collisions led him to formulate a simple law relating to the speeds before and after a direct collision between two bodies, called *Newton's law of impact*.

$$\frac{\text{speed of separation}}{\text{speed of approach}} = \text{constant}$$

This can also be written as

$$\text{speed of separation} = \text{constant} \times \text{speed of approach}$$

This constant is called the *coefficient of restitution* and is conventionally denoted by the letter e. For two particular surfaces, e is a constant between 0 and 1. It does not have units, being the ratio of two speeds.

For very bouncy balls, e is close to 1, and for balls that do not bounce, e is close to 0. A collision for which $e = 1$ is called perfectly elastic, and a collision for which $e = 0$ is called perfectly inelastic. Some values are given in the table.

Surfaces in contact	e
Two glass marbles	0.95
Two snooker balls	0.80
Slow squash ball and racquet	0.25
Fast squash ball and racquet	0.32
Steel and steel	0.55
Golf ball and club	0.75
Snooker ball and classroom table	0.56
Tennis ball and classroom table	0.72
Tennis ball and carpet	0.71
Table tennis ball and classroom table	0.77
Table tennis ball and carpet	0.18

DIRECT IMPACT WITH A FIXED SURFACE

The value of e for the ball that you used in the experiment is given by $\frac{v_s}{v_a}$, and you should have found that this had approximately the same value each time for any particular ball.

When a moving object hits a fixed surface which is perpendicular to its motion it rebounds in the opposite direction. If the speed of approach is v_a and the speed of separation is v_s Newton's law of impact gives

$$\frac{v_s}{v_a} = e$$

$$\Rightarrow \quad v_s = ev_a$$

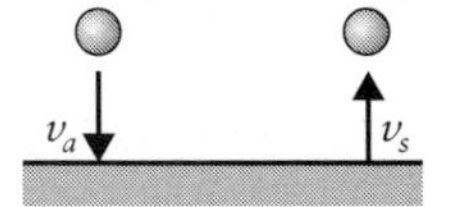

FIGURE 5.6

COLLISIONS BETWEEN TWO BODIES MOVING IN THE SAME STRAIGHT LINE

Figure 5.7 shows two objects that collide while moving along a straight line. Object A is catching up with B, and after the collision either B moves away from A or they continue together.

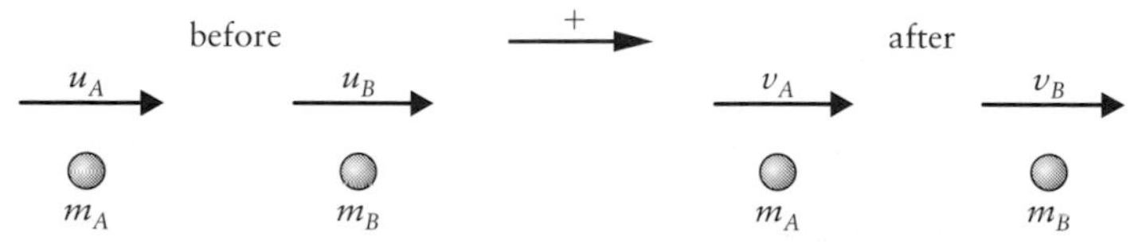

FIGURE 5.7

Speed of approach $u_A - u_B$. ($u_A > u_B$ for the collision to occur)

Speed of separation $v_B - v_A$. ($v_B > v_A$ as B moves away from A)

By Newton's law of impact

$$\text{speed of separation} = e \times \text{speed of approach}$$

$$\Rightarrow \quad v_B - v_A = e(u_A - u_B) \qquad ①$$

A second equation relating the velocities follows from the law of conservation of momentum in the positive direction (→)

$$\text{momentum after collision} = \text{momentum before collision}$$

$$m_A v_A + m_B v_B = m_A u_A + m_B u_B \qquad ②$$

These two equations, ① and ②, allow you to calculate the final velocities, v_A and v_B, after any collision as shown in the next two examples.

EXAMPLE 5.3

A direct collision takes place between two snooker balls. The cue ball travelling at $2\,\text{ms}^{-1}$ hits a stationary red ball. After the collision the red ball moves in the direction in which the cue ball was moving before the collision. Assume that the balls have equal mass, and that the coefficient of restitution between the two balls is 0.6. Predict the velocities of the two balls after the collision.

Solution

Let the mass of each ball be m, and call the (white) cue ball 'W' and the red ball 'R'. The situation is summarised in figure 5.8.

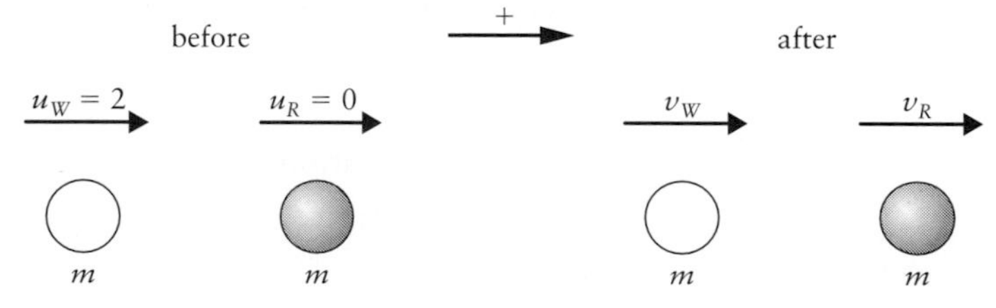

FIGURE 5.8

Speed of approach = 2 − 0 = 2 Speed of separation = $v_R - v_W$

By Newton's law of impact

$$\text{speed of separation} = e \times \text{speed of approach}$$
$$\Rightarrow \quad v_R - v_W = 0.6 \times 2$$
$$\Rightarrow \quad v_R - v_W = 1.2 \qquad ①$$

Conservation of momentum gives

$$mv_W + mv_R = mu_W + mu_R$$

Dividing through by m, and substituting $u_W = 2$, $u_R = 0$, this becomes

$$v_W + v_R = 2 \qquad ②$$

Adding equations ① and ② gives $2v_R = 3.2$,

so $v_R = 1.6$, and from equation ②, $v_W = 0.4$.

After the collision both balls move forward, the red ball at a speed of $1.6\,\text{ms}^{-1}$ and the cue ball at a speed of $0.4\,\text{ms}^{-1}$.

MECHANICAL ENERGY DURING IMPACTS

EXAMPLE 5.4

An object A of mass m moving with speed $2u$ hits an object B of mass $2m$ moving with speed u in the opposite direction from A.

(a) Show that the ratio of speeds remains unchanged whatever the value of e.
(b) Find the loss of kinetic energy in terms of m, u and e.

Solution **(a)** Let the velocities of A and B after the collision be v_A and v_B respectively.

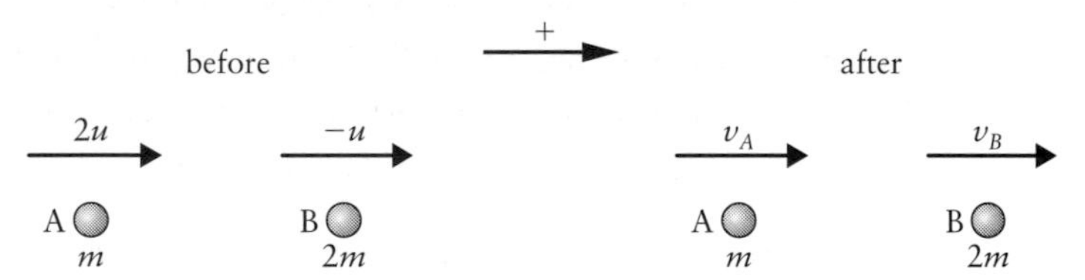

FIGURE 5.9

$$\text{Speed of approach} = 2u - (-u) = 3u$$
$$\text{Speed of separation} = v_B - v_A$$

Using Newton's law of impact

$$\text{speed of separation} = e \times \text{speed of approach}$$
$$\Rightarrow \quad v_B - v_A = e \times 3u \qquad ①$$

Conservation of momentum gives

$$mv_A + 2mv_B = m(2u) + 2m(-u)$$

Dividing by m gives

$$v_A + 2v_B = 0 \quad ②$$

Equation ① is $v_B - v_A = 3eu$

Adding ① and ② $3v_B = 3eu$

$$v_B = eu$$

From ② $v_A = -2eu$

The ratio of speeds was initially $2u : u$ and finally $2eu : eu$ so the ratio of speeds is unchanged at 2 : 1 (providing $e \neq 0$).

(b) Initial K.E. of A $\frac{1}{2}m \times (2u)^2 = 2mu^2$

Initial K.E. of B $\frac{1}{2}(2m) \times u^2 = mu^2$

Total K.E. before impact $= 3mu^2$

Final K.E. of A $\frac{1}{2}m \times 4e^2u^2 = 2me^2u^2$

Final K.E. of B $\frac{1}{2}(2m) \times e^2u^2 = me^2u^2$

Total K.E. after impact $= 3me^2u^2$

Loss of K.E. $= 3mu^2(1 - e^2)$.

Note In this case, A and B lose *all* their energy when e = 0, but this is not true in general. Only when e = 1 is there *no* loss in K.E. Kinetic energy is lost in any collision in which the coefficient of restitution is not equal to 1.

EXERCISE 5B

You will find it helpful to draw diagrams when answering these questions.

1 In each of the situations shown below, find the unknown quantity, either the initial speed u, the final speed v or the coefficient of restitution e.

(a)

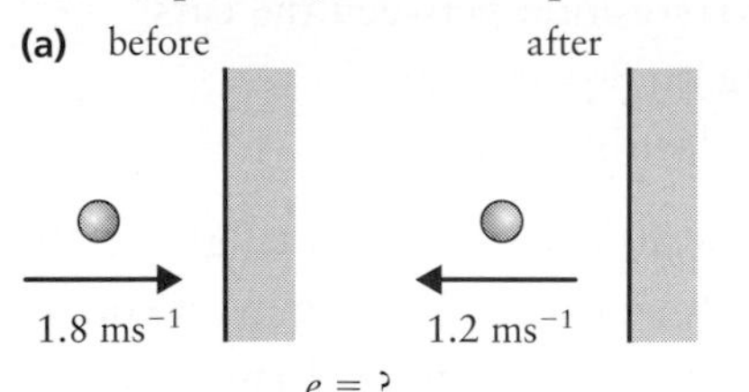

(b)

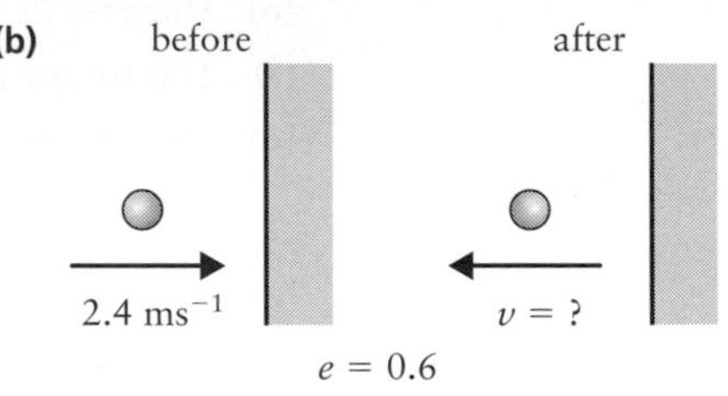

(c)

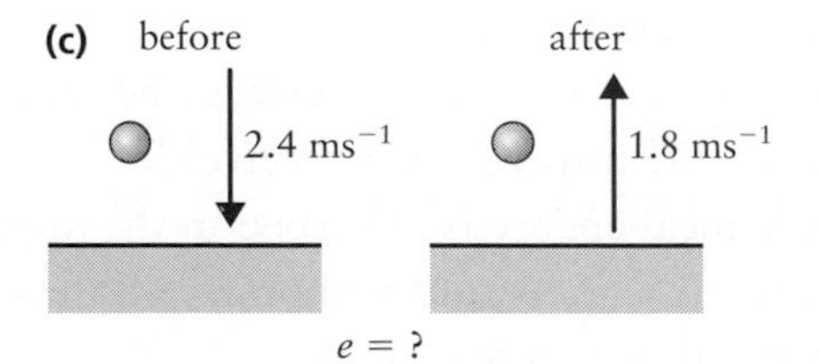

(d)

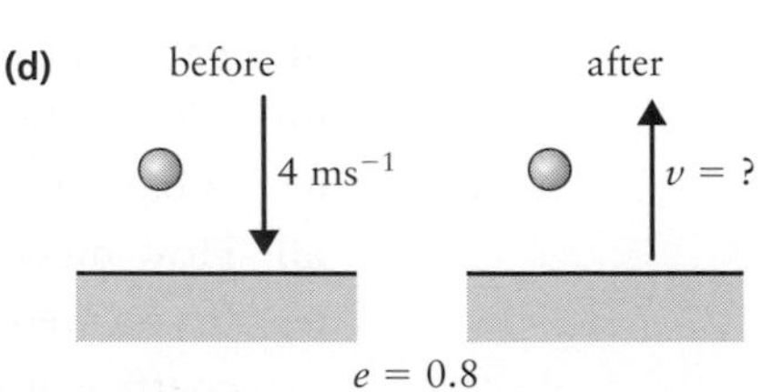

2 Find the coefficient of restitution in the following situations.

(a) A football hits the goalpost at $10\,\mathrm{ms}^{-1}$ and rebounds in the opposite direction with speed $3\,\mathrm{ms}^{-1}$.

(b) A beanbag is thrown against the wall with speed $5\,\mathrm{ms}^{-1}$ and falls straight down to the ground.

(c) A superball is dropped on to the ground, landing with speed $8\,\mathrm{ms}^{-1}$ and rebounding with speed $7.6\,\mathrm{ms}^{-1}$.

(d) A photon approaches a mirror along a line normal to its surface with speed $3 \times 10^8\,\mathrm{ms}^{-1}$ and leaves it along the same line with speed $3 \times 10^8\,\mathrm{ms}^{-1}$.

3 A tennis ball of mass 60 g is hit against a practice wall. At the moment of impact it is travelling horizontally with speed $15\,\mathrm{ms}^{-1}$. Just after the impact its speed is $12\,\mathrm{ms}^{-1}$, also horizontally. Find

(a) the coefficient of restitution between the ball and the wall

(b) the impulse acting on the ball

(c) the loss of kinetic energy during the impact.

4 A ball of mass 80 g is dropped from a height of 1 m on to a level floor and bounces back to a height of 0.81 m. Find

(a) the speed of the ball just before it hits the floor

(b) the speed of the ball just after it has hit the floor

(c) the coefficient of restitution

(d) the change in the kinetic energy of the ball from just before it hits the floor to just after it leaves the floor

(e) the change in the potential energy of the ball from the moment when it was dropped to the moment when it reaches the top of its first bounce

(f) the height of the ball's next bounce.

5 Two children drive dodgems straight at each other, and collide head-on. Both dodgems have the same mass (including their drivers) of 150 kg. Isobel is driving at $3\,\mathrm{ms}^{-1}$, Stuart at $2\,\mathrm{ms}^{-1}$. After the collision Isobel is stationary. Find

(a) Stuart's velocity after the collision

(b) the coefficient of restitution between the cars

(c) the impulse acting on Stuart's car

(d) the kinetic energy lost in the collision.

6 A trapeze artist of mass 50 kg falls from a height of 20 m into a safety net.

(a) Find the speed with which she hits the net. (You may ignore air resistance and should take the value of g to be $10\,\mathrm{ms}^{-2}$.)

Her speed on leaving the net is $15\,\mathrm{ms}^{-1}$.

(b) What is the coefficient of restitution between her and the net?

(c) What impulse does the trapeze artist receive?

(d) How much mechanical energy is absorbed in the impact?

(e) If you were a trapeze artist would you prefer a safety net with a high coefficient of restitution or a low one?

7 In each of the situations **(a)–(f)**, a collision is about to occur. Masses are given in kilograms, speeds are in metres per second.

(i) Draw diagrams showing the situations before and after the impact, indicating the values of any velocities which you know, and the symbols you are using for those which you do not know.

(ii) Use the equations corresponding to the law of conservation of momentum and to Newton's law of impact to find the final velocities.

(iii) Find the loss of kinetic energy during the collision.

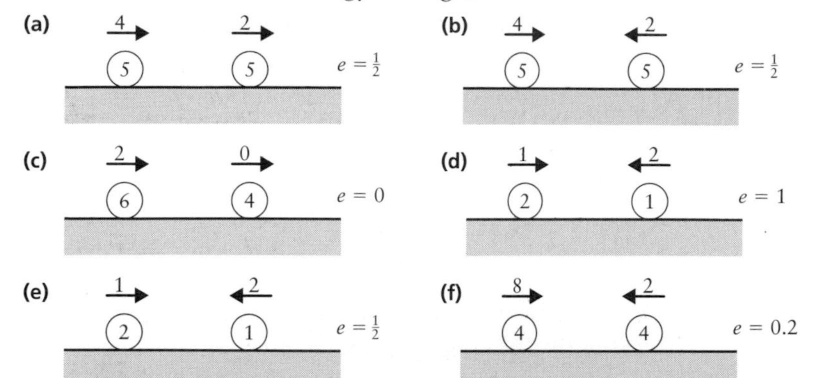

8 Two spheres of equal mass, m, are travelling towards each other along the same straight line when they collide. Both have speed v just before the collision and the coefficient of restitution between them is e. Your answers should be given in terms of m, v and e.

(a) Draw diagrams to show the situation before and after the collision.

(b) Find the velocities of the spheres after the collision.

(c) Show that the kinetic energy lost in the collision is given by $mv^2(1 - e^2)$.

(d) Use the result in part **(c)** to show that e cannot have a value greater than 1.

9 A sphere, A, of mass 40 g travels with speed $3\,\text{ms}^{-1}$ along the line of centres towards another sphere, B, of the same size but of mass 60 g which is initially stationary. After the collision sphere A is stationary.

(a) Draw diagrams to illustrate the situation before and after the collision.

(b) Find the velocity of sphere B after the collision.

(c) Find the coefficient of restitution between the spheres.

This situation may be generalised so that sphere A, of mass m_1, moving with speed u_1 along the line of centres collide with the stationary sphere B, of mass m_2. After the collision sphere A is stationary. The coefficient of restitution between the spheres is e.

(d) Show that $e = \dfrac{m_1}{m_2}$.

(e) What does this tell you about the relative mass of the two spheres?

(f) Find the ratio of the kinetic energy after impact to that before impact and write it in terms of e. What must be the value of e if kinetic energy is to be conserved when the spheres collide?

10 The coefficient of restitution between a ball and the floor is e. The ball is dropped from a height h. Air resistance may be neglected, and your answers should be given in terms of e, h, g and n, the number of bounces.

(a) Find the time it takes the ball to reach the ground and its speed when it arrives there.

(b) Find the ball's height at the top of its first bounce.

(c) Find the height of the ball at the top of its nth bounce.

(d) Find the time that has elapsed when the ball hits the ground for the second time, and for the nth time.

(e) Show that according to this model the ball comes to rest within a finite time having completed an infinite number of bounces.

(f) What distance does the ball travel before coming to rest?

SUCCESSIVE IMPACTS

When there are several spheres in a straight line there will be several impacts. You deal with these one at a time exactly as above. You should take care over the signs, particularly if you are deciding whether there will be a collision. Diagrams are very helpful.

The next two examples show this.

EXAMPLE 5.5

Three uniform smooth spheres A, B and C are of the same size and lying in a straight line in that order on a smooth table. Their masses are $4m$, $2m$ and $3m$ respectively. Initially B and C are at rest, and A is moving directly towards B with velocity $6u$. The coefficient of restitution between A and B is $\frac{1}{2}$ and the coefficient of restitution between B and C is $\frac{1}{3}$.

(a) Find the velocities of A and B after the first collision.

(b) Find the velocities of B and C after the second collision.

(c) Will there be any further collisions? Explain your answer.

Solution **(a)** Let the velocities of A and B after the collision be v_A and v_B respectively.

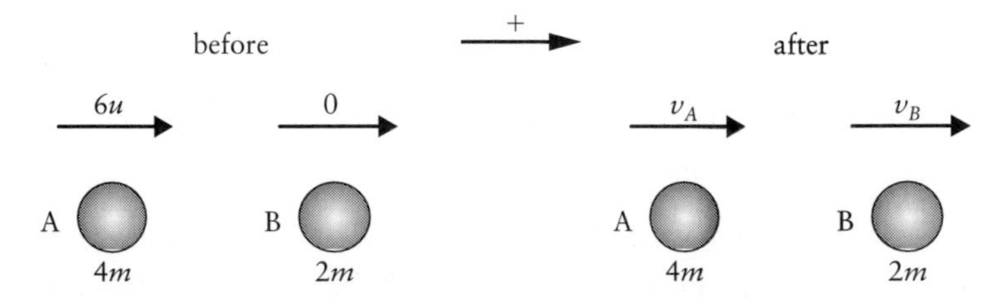

FIGURE 5.10

Using Newton's law of impact

$$\text{speed of separation} = e \times \text{speed of approach}$$

$$\Rightarrow \quad v_B - v_A = \tfrac{1}{2} \times 6u = 3u \qquad ①$$

Conservation of momentum gives

$$4mv_A + 2mv_B = 4m \times 6u = 24mu$$

Dividing by $2m$ gives $\quad 2v_A + v_B = 12u \qquad ②$

Equation ① gives $\quad -v_A + v_B = 3u$

Subtracting $\quad 3v_A = 9u$

Giving $\quad v_A = 3u$

$\quad v_B = 6u$

(b) Let the velocities of B and C after the collision be w_B and w_C respectively.

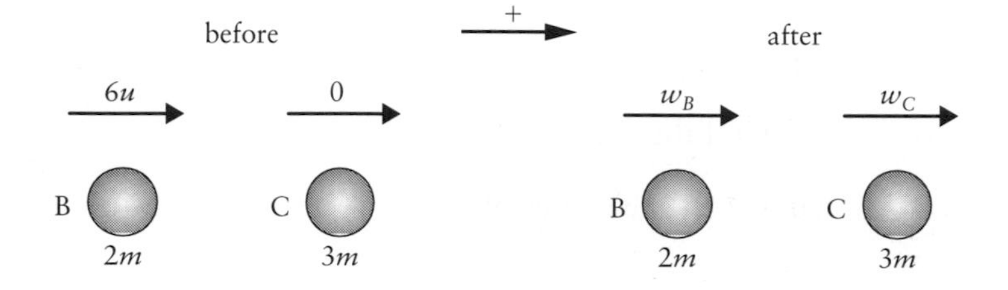

FIGURE 5.11

$$\text{Speed of approach} = 6u$$
$$\text{Speed of separation} = w_C - w_B$$

Using Newton's law of impact

$$\text{speed of separation} = e \times \text{speed of approach}$$

$$\Rightarrow \quad w_C - w_B = \tfrac{1}{3} \times 6u = 2u \qquad ③$$

Conservation of momentum gives

$$2mw_B + 3mw_C = 2m \times 6u = 12mu$$

Dividing by m gives $\quad 2w_B + 3w_C = 12u \qquad ④$

2 × equation ③ gives $\quad -2w_B + 2w_C = 4u$

Adding $\quad 5w_C = 16u$

Giving $\quad w_C = 3.2u$

$\quad w_B = 1.2u$

(c) The diagram shows the velocities of the three spheres after the second collision.

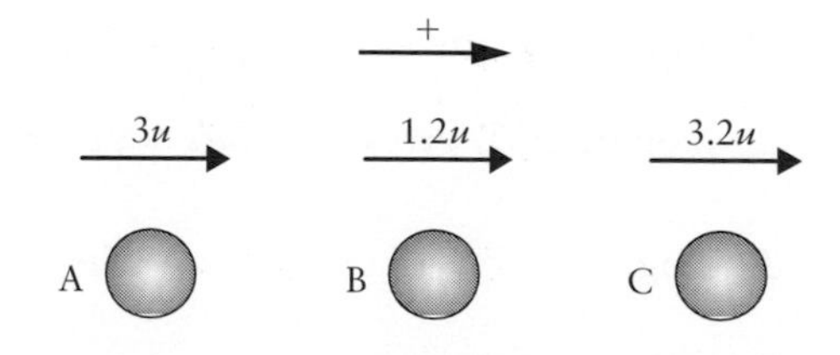

FIGURE 5.12

As A is moving towards B with a greater velocity than that of B, there will be another collision between A and B. It is possible that B might then strike C again, as its velocity will increase in that collision.

EXAMPLE 5.6

Two uniform smooth spheres A and B are of equal size and masses $2m$ and $3m$ respectively. They lie on a line which is perpendicular to a vertical wall. B is closer to the wall and is a distance of d from the wall. A is moving directly towards B with velocity u. The coefficient of restitution between A and B is $\frac{1}{3}$. The coefficient of restitution between B and the wall is e.

(a) Find the velocities of A and B after the first collision.

B then bounces off the wall and returns to collide with A again. The second collision between A and B takes place at a distance $\frac{1}{2}d$ from the wall.

(b) Find the value of e.

Solution **(a)** Let the velocities of A and B after the collision be v_A and v_B respectively.

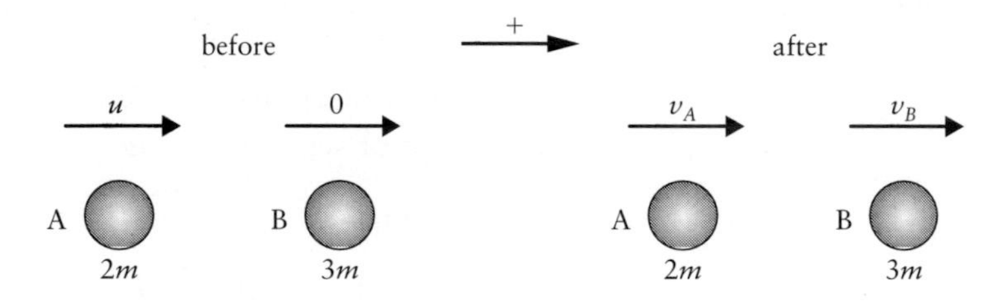

FIGURE 5.13

$$\text{Speed of approach} = u$$
$$\text{Speed of separation} = v_B - v_A$$

Using Newton's law of impact

$$\text{speed of separation} = e \times \text{speed of approach}$$

$$\Rightarrow \qquad v_B - v_A = \frac{1}{3} \times u = \frac{u}{3} \qquad ①$$

Conservation of momentum gives

$$2mv_A + 3mv_B = 2m \times u = 2mu$$

Dividing by m gives $\qquad 2v_A + 3v_B = 2u \qquad ②$

$2 \times$ equation ① gives $\qquad -2v_A + 2v_B = \dfrac{2u}{3}$

Adding $\qquad 5v_B = \dfrac{8u}{3}$

Giving $$v_B = \frac{8u}{15}$$

$$v_A = \frac{u}{5}$$

(b) B will hit the wall at time $\frac{15d}{8u}$.

During this time A moves a distance $\frac{u}{5} \times \frac{15d}{8u} = \frac{3d}{8}$.

The spheres are $\frac{5d}{8}$ apart as B bounces off the wall, with speed $e \times \frac{8u}{15}$.

The diagram shows the situation.

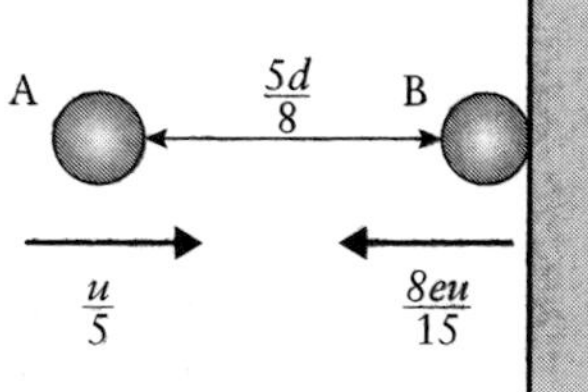

FIGURE 5.14

The spheres are approaching each other at a relative speed of $\left(\frac{u}{5} + \frac{8eu}{15}\right)$

$$= \frac{u}{15}(3 + 8e)$$

$$\text{Time} = \frac{\text{distance}}{\text{speed}} \text{ so}$$

the time to the next collision $= \dfrac{\frac{5d}{8}}{\frac{u}{15}(3 + 8e)}$

$$= \frac{15 \times 5d}{8u(3 + 8e)}$$

The distance B travels $= \dfrac{15 \times 5d}{8u(3 + 8e)} \times \dfrac{8eu}{15} = \dfrac{5de}{3 + 8e}$

$$\frac{5de}{3 + 8e} = \frac{d}{2}$$

Cross-multiplying $10de = (3 + 8e)d$

Dividing by d $10e = 3 + 8e$

Giving $e = \frac{2}{3}$

Note There is an alternative method for part **(b)** based on the ratios of speeds which does not find the time. This would also be satisfactory.

Exercise 5C

1 Three identical spheres are lying in the same straight line. The coefficient of restitution between any pair of spheres is $\frac{1}{2}$. The left-hand ball is given speed $2\,\mathrm{ms}^{-1}$ towards the other two. What are the final velocities of all three, when no more collisions can occur?

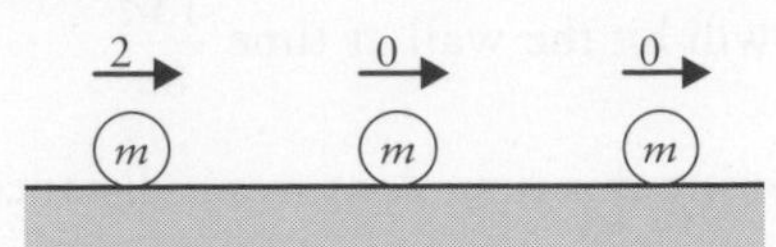

2 The diagram shows two snooker balls and one edge cushion. The coefficient of restitution between the balls and the cushion is 0.5 and that between the balls is 0.75. Ball A (the cue ball) is hit directly towards the stationary ball B with speed $8\,\mathrm{ms}^{-1}$. Find the speeds and directions of the two balls after their second impact with each other.

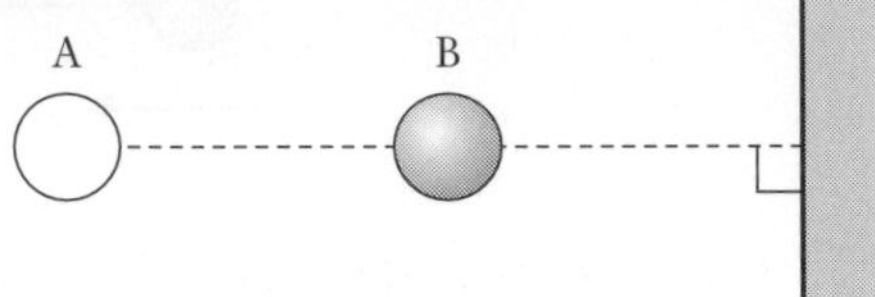

3 Three uniform smooth spheres A, B and C are of the same size and lying in a straight line in that order on a smooth table. Their masses are $4m$, $5m$ and $8m$ respectively. Initially B and C are at rest, and A is moving directly towards B with velocity u. The coefficient of restitution between A and B is $\frac{1}{4}$ and the coefficient of restitution between B and C is $\frac{3}{4}$.

(a) Find the velocities of A and B after the first collision.

(b) Find the velocities of B and C after the second collision.

(c) Will there be any further collisions? Explain your answer.

4 The diagram shows two equal-sized snooker balls each of mass m kg and one edge cushion. The coefficient of restitution between the balls and the cushion is 0.5 and that between the balls is 0.7. Ball A (the cue ball) is hit directly towards the stationary ball B with speed $u\,\mathrm{ms}^{-1}$.

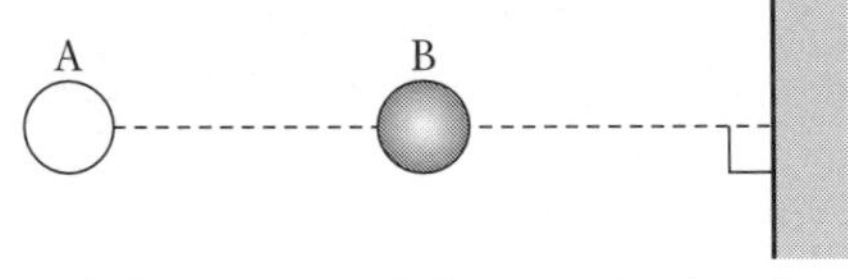

(a) Find the speeds and directions of the two balls after their first impact with each other.

Ball B now collides directly with the cushion.

(b) Find the impulse given to ball B by the cushion.

Balls A and B now collide directly again.

(c) Find the velocities of the two balls after their second collision.

(d) Find the kinetic energy lost in the series of collisions.

(e) Check that momentum has been conserved through the whole series of collisions.

5 Three uniform smooth spheres A, B and C are of the same size and lying in a straight line in that order on a smooth table. Their masses are $2m$, $4m$ and $5m$ respectively. Initially B and C are at rest, and A is moving directly towards B with velocity u. The coefficient of restitution between any two spheres is e.

A collides directly with B and then B collides directly with C. After the second collision B is at rest.

(a) Find e.

(b) Find the velocities of A and C after both collisions.

EXERCISE 5D

Examination-style questions

1 A ball of mass 0.3 kg is moving with a velocity of $2\mathbf{i} + 4\mathbf{j}\,\text{ms}^{-1}$ when it is given an impulse of $-0.9\mathbf{i} + 0.6\mathbf{j}\,\text{Ns}$.

(a) Find the new velocity of the ball.

(b) Find the angle through which the ball is deflected by the impulse, giving your answer in degrees to 1 decimal place.

2 A ball of mass 0.15 kg is moving with a velocity of $20\mathbf{i} + 4\mathbf{j}\,\text{ms}^{-1}$ when it is struck by a bat. After the collision, the ball has a velocity of $-10\mathbf{i} + 8\mathbf{j}\,\text{ms}^{-1}$.

(a) Find the impulse given to the ball by the bat.

(b) Find the change in the direction of motion of the ball, giving your answer in degrees to 1 decimal place.

(c) Find the change in kinetic energy of the ball, stating whether it is a gain or a loss.

3 During a hailstorm, hailstones are hitting a window pane at a mean rate of 100 per second. Each one may be modelled as a mass of 0.5 g which hits the window with velocity $5\mathbf{i} - 3\mathbf{j}\,\text{ms}^{-1}$ and rebounds with velocity $-2\mathbf{i} - 3\mathbf{j}\,\text{ms}^{-1}$. Find the mean force of the hailstones on the window.

4 Two uniform smooth spheres, A and B, are moving towards each other on a smooth horizontal table. A has mass $3m$ and speed $2u$ and B has mass $4m$ and speed u. A and B collide directly, with coefficient of restitution e.

(a) If the velocity of A is reversed in the collision, show that $e > \frac{1}{6}$.

(b) Show that, whatever the value of e, the velocity of B after the collision cannot exceed $\frac{11u}{7}$.

(c) If $e = \frac{1}{3}$, find the kinetic energy lost in the collision.

5 Two uniform smooth spheres, A and B, are moving in the same straight line on a smooth horizontal table. A has mass $3m$ and velocity $6u$ towards B, while B has mass $2m$ and velocity $2u$ away from A. A and B collide directly. The coefficient of restitution between them is e.

(a) Show that whatever the value of e the direction of motion of A cannot be reversed.

(b) After the collision the velocity of A is half the velocity of B. Find the value of e.

(c) Find the kinetic energy lost in the collision, giving your answer in terms of m and u.

6 Two uniform smooth spheres, A and B, are moving on a smooth horizontal table. A has mass $3m$ and velocity u towards B, while B has mass $2m$ and is initially stationary. A and B collide directly. The coefficient of restitution between them is e. After the collision, half of the initial kinetic energy has been lost. Find the value of e.

7 Three uniform smooth spheres A, B and C are of equal radius and lie with their centres in a straight line in that order on a smooth table. They are of masses $3m$, $2m$ and $4m$ respectively. The coefficient of restitution between A and B is $\frac{2}{3}$, while that between B and C is e. A is projected directly towards B with velocity u and collides with B.

(a) Show that after the collision B has velocity u, and find the velocity of A.

B then collides directly with C.

(b) Show that if $e \neq 0$ there will be a second collision between A and B.

8 Three uniform smooth spheres A, B and C are of equal radius and lie with their centres in a straight line in that order on a smooth table. They are of masses $4m$, $6m$ and $8m$ respectively. The coefficient of restitution between A and B is e, while that between B and C is 0.9. A is projected directly towards B with velocity u and collides with B.

(a) Show that the direction of the velocity of A is reversed if $e > \frac{2}{3}$.

B then collides directly with C.

(b) Find the velocity of B after this collision.

(c) Show that there will be another collision between A and B if $e < \frac{79}{99}$.

9 Two smooth spheres P and Q are of equal radius and lie on a smooth table with their centres making a line which is perpendicular to a vertical wall. Q lies between P and the wall. P is of mass m and Q is of mass $2m$. The coefficient of restitution between P and Q is $\frac{2}{3}$, and that between Q and the wall is e. P is projected directly towards Q and collides with it.

(a) Find the velocities of P and Q after this collision.

Q then collides with the wall.

(b) Show that there will be a second collision between P and Q if $e > \frac{1}{5}$.

(c) If $e = 0.4$, find an expression for the total kinetic energy lost after the second collision between P and Q.

10 Two smooth spheres X and Y are of equal radius and lie on a smooth table with their centres making a line which is perpendicular to a vertical wall. Y lies between X and the wall. X is of mass $2m$ and Y is of mass $3m$. The coefficient of restitution between X and Y is $\frac{1}{2}$, and that between Y and the wall is $\frac{1}{3}$. X is projected directly towards Y with initial velocity u and collides directly with it. Find the velocities of X and Y after all collisions have finished.

11 A tennis ball of mass 0.2 kg is moving with velocity $(-10\mathbf{i})\,\mathrm{ms}^{-1}$ when it is struck by a tennis racket. Immediately after being struck, the ball has velocity $(15\mathbf{i} + 15\mathbf{j})\,\mathrm{ms}^{-1}$. Find

(a) the magnitude of the impulse exerted by the racket on the ball.

(b) the angle to the nearest degree, between the vector **i** and the impulse exerted by the racket.

(c) the kinetic energy gained by the ball as a result of being struck.

[Edexcel]

12 A uniform sphere A of mass m is moving with speed u on a smooth horizontal table when it collides directly with another uniform sphere B of mass $2m$ which is at rest on the table. The spheres are of equal radius and the coefficient of restitution between them is e. The direction of motion of A is unchanged by the collision.

(a) Find the speeds of A and B immediately after the collision.

(b) Find the range of possible values of e.

After being struck by A, the sphere B collides directly with another sphere C, of mass $4m$ and of the same size as B. The sphere C is at rest on the table immediately before being struck by B. The coefficient of restitution between B and C is also e.

(c) Show that, after B has struck C, there will be a further collision between A and B.

[Edexcel]

13

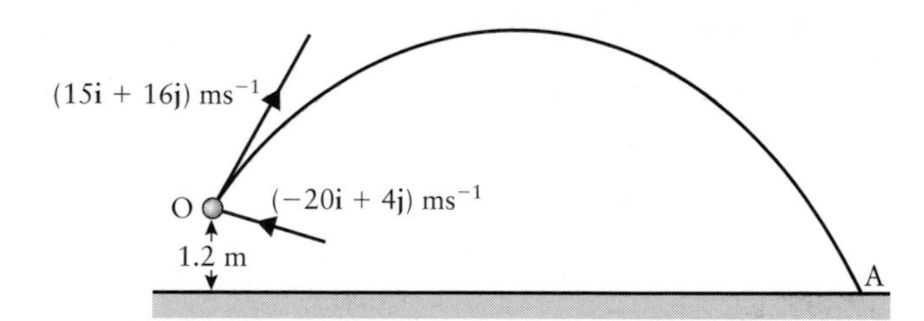

A ball B of mass 0.4 kg is struck by a bat at a point O which is 1.2 m above horizontal ground. The unit vectors **i** and **j** are respectively horizontal and vertical. Immediately before being struck, B has velocity $(-20\mathbf{i} + 4\mathbf{j})\,\mathrm{ms}^{-1}$. Immediately after being struck it has velocity $(15\mathbf{i} + 16\mathbf{j})\,\mathrm{ms}^{-1}$.

After B has been struck, it moves freely under gravity and strikes the ground at the point A, as shown in the diagram. The ball is modelled as a particle.

(a) Calculate the magnitude of the impulse exerted by the bat on B.

(b) By using the principle of conservation of energy, or otherwise, find the speed of B when it reaches A.

(c) Calculate the angle which the velocity of B makes with the ground when B reaches A.

(d) State two additional physical factors which could be taken into account in a refinement of the model of the situation which would make it more realistic.

[Edexcel]

KEY POINTS

1. The impulse from a force **F** is given by **F**t where t is the time for which the force acts. Impulse is a vector quantity.
2. The momentum of a body of mass m travelling with velocity **v** is given by m**v**. Momentum is a vector quantity.
3. The impulse–momentum equation is

   impulse = final momentum – initial momentum
4. The law of conservation of momentum states that, when no external forces are acting on a system, the total momentum of the system is constant. Since momentum is a vector quantity this applies to the magnitude of the momentum in any direction.
5. Coefficient of restitution $e = \dfrac{\text{speed of separation}}{\text{speed of approach}}$

   speed of separation = e × speed of approach
6. $0 \leqslant e \leqslant 1$
7. Newton's law of restitution applies perpendicular to a smooth plane.

Chapter six

STATICS OF RIGID BODIES

Give me a firm place to stand and I will move the earth.

Archimedes

EXPERIMENT

Set up the apparatus shown in figure 6.1 below and experiment with two or more weights in different positions.

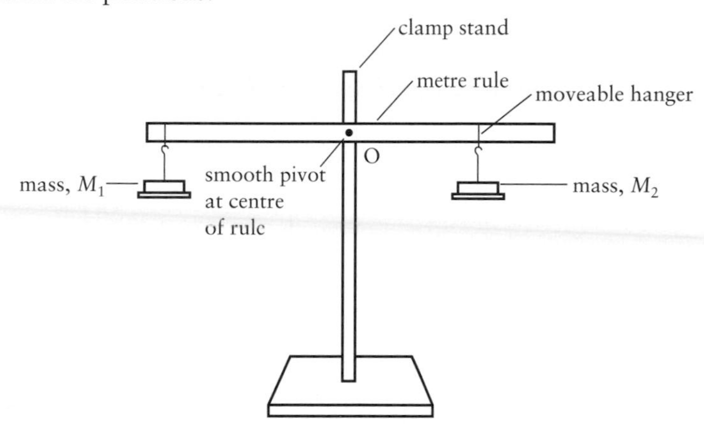

FIGURE 6.1

Record your results in a table showing weights, distances from O and moments about O.

Two masses are suspended from the rule in such a way that the rule balances in a horizontal position. What happens when the rule is then moved to an inclined position and released?

Now attach a pulley as in figure 6.2. Start with equal weights and measure *d* and *l*. Then try different weights and pulley positions.

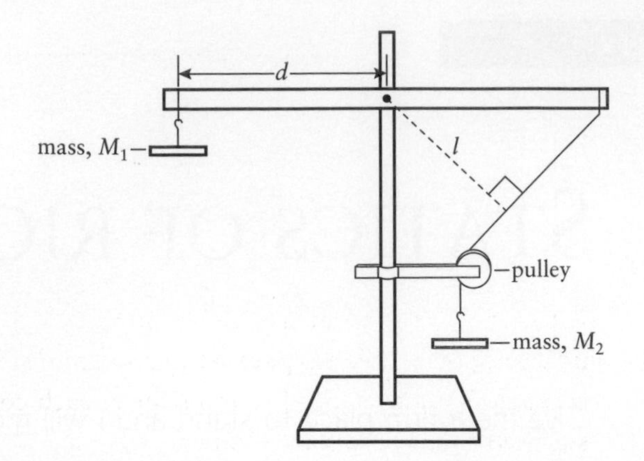

Figure 6.2

The moment of a force which acts at an angle

From the experiment you will have seen that the moment of a force about the pivot depends on the *perpendicular distance* from the pivot to the line of the force.

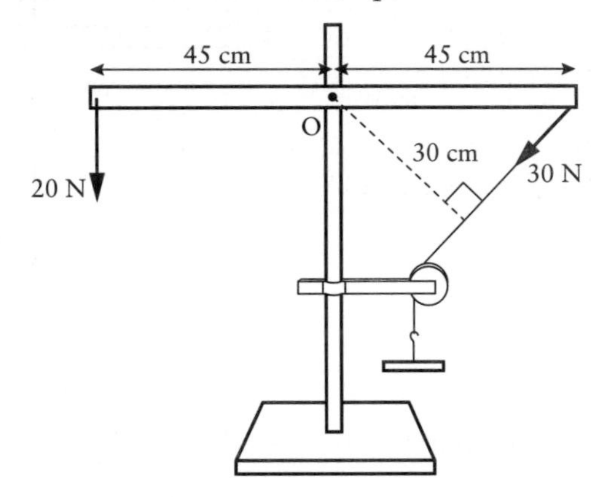

Figure 6.3

In figure 6.3, where the system remains at rest, the moment about O of the 20 N force is $20 \times 0.45 = 9$ Nm. The moment about O of the 30 N force is $-30 \times 0.3 = -9$ Nm. The system is in equilibrium even through unequal forces act at equal distances from the pivot.

The magnitude of the moment of the force F about O in figure 6.4 is given by

$$F \times l = Fd\sin\alpha$$

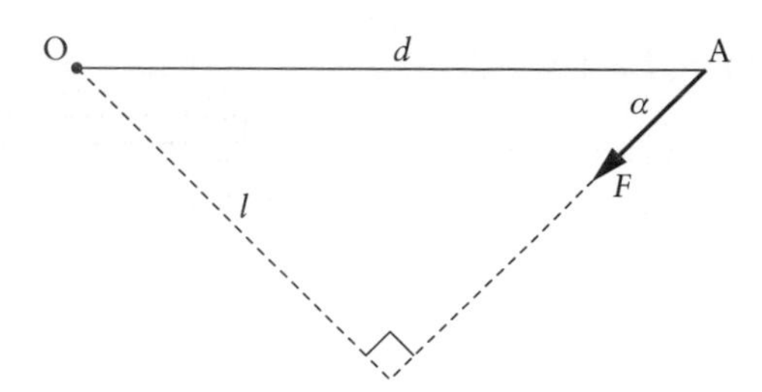

FIGURE 6.4

Alternatively the moment can be found by noting that the force F can be resolved into components $F\cos\alpha$ parallel to AO and $F\sin\alpha$ perpendicular to AO, both acting through A (figure 6.5). The moment of each component can be found and then summed to give the total moment.

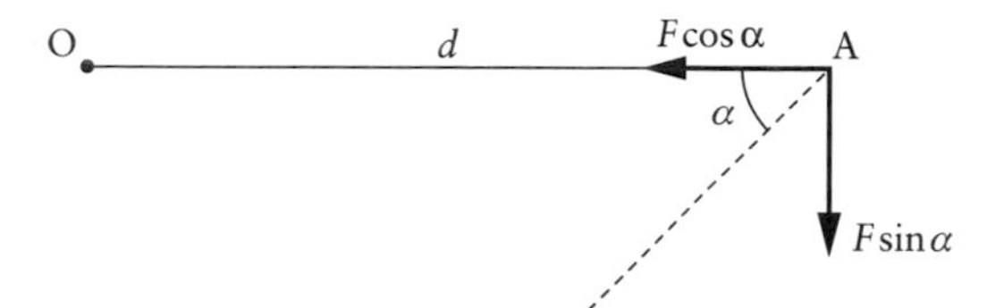

FIGURE 6.5

The moment of the component along AO is zero because it acts through O. The magnitude of the moment of the perpendicular component is $F\sin\alpha \times d$ so the total moment is $Fd\sin\alpha$, as expected.

EXAMPLE 6.1

A force of 40 N is exerted on a rod as shown. Find the moment of the force about the point marked O.

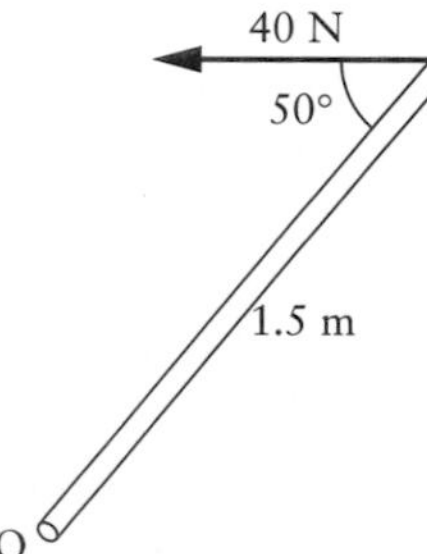

FIGURE 6.6

Solution In order to calculate the moment, the perpendicular distance between O and the line of action of the force must be found. This is shown on the diagram.

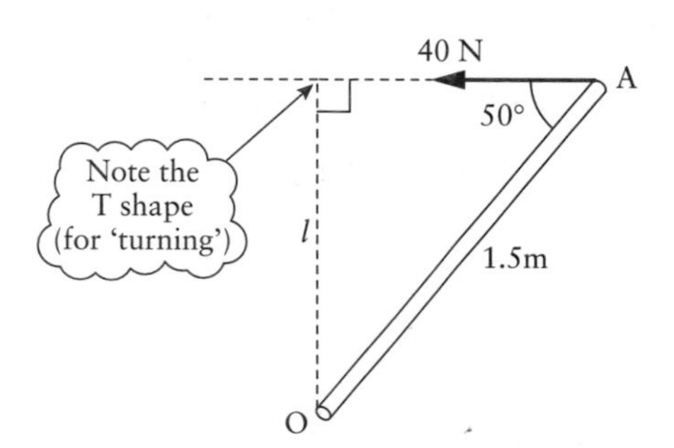

FIGURE 6.7

Here $l = 1.5 \times \sin 50°$.

So the moment about O is

$$F \times l = 40 \times (1.5 \times \sin 50°)$$
$$= 46.0\,\text{Nm}$$

Alternatively you can resolve the 40 N force into components as in figure 6.8.

The component of the force parallel to AO is 40cos50° N. The component perpendicular to AO is 40sin50° (or 40cos40°) N.
So the moment about O is
$40\sin 50° \times 1.5 = 60\cos 40°$
$= 46.0\,\text{Nm}$ as before

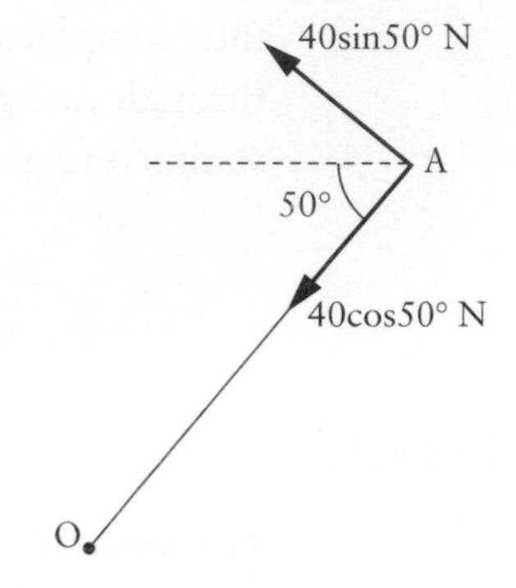

FIGURE 6.8

EXAMPLE 6.2

A sign outside a pub is attached to a light rod of length 1 m which is freely hinged to the wall and supported in a vertical plane by a light string as in the diagram. The sign is assumed to be a uniform rectangle of mass 10 kg. The angle of the string to the horizontal is 25°.

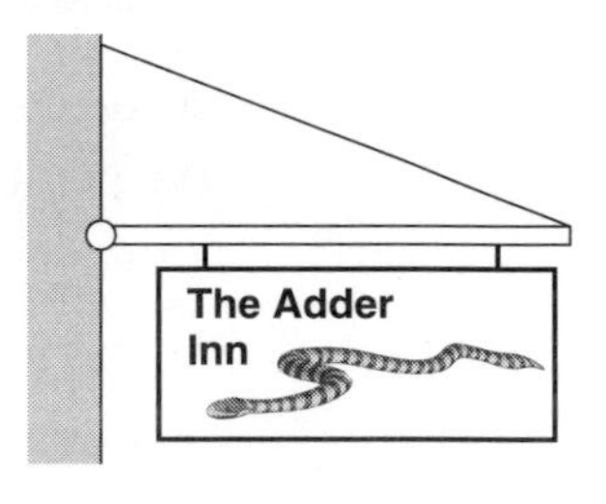

FIGURE 6.9

(a) Find the tension in the string.
(b) Find the magnitude and direction of the reaction force of the hinge on the sign.

Solution

(a) The diagram shows the forces acting on the rod, where R_H and R_V are the magnitudes of the horizontal and vertical components of the reaction **R** on the rod at the wall.

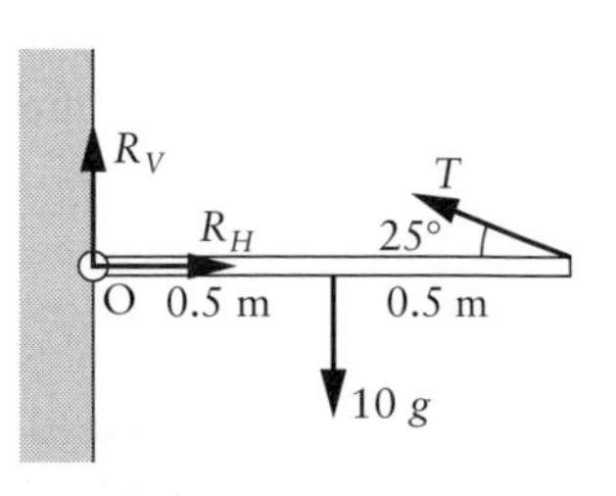

FIGURE 6.10

Taking moments about O

$$R \times 0 - 10g \times 0.5 + T\sin 25° \times 1 = 0$$
$$\Rightarrow T\sin 25° = 5g$$
$$T = 116$$

The tension is 116 N.

(b) You can resolve to find the reaction at the wall.

Horizontally $\qquad R_H = T\cos 25°$

$\qquad\qquad \Rightarrow R_H = 105$

Vertically $\qquad R_V + T\sin 25° = 10g$

$\qquad\qquad \Rightarrow R_V = 10g - 5g = 49$

$$R = \sqrt{105^2 + 49^2}$$
$$= 116$$

$$\theta = \arctan\left(\frac{49}{105}\right) = 25°$$

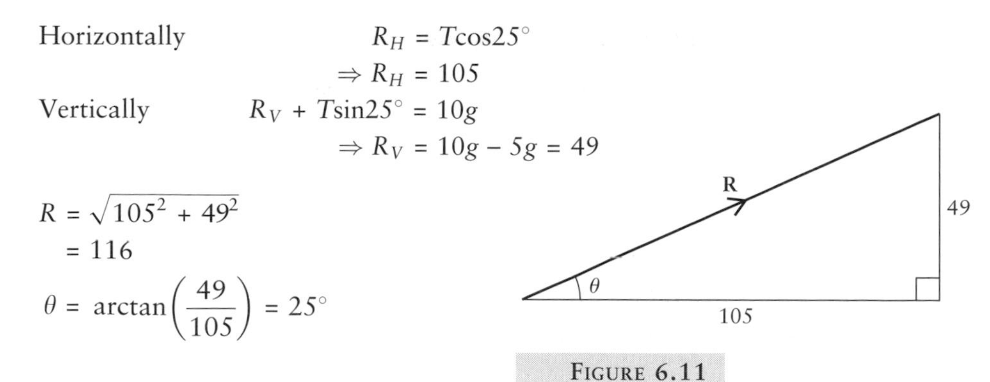

FIGURE 6.11

The reaction at the hinge has magnitude 116 N and acts at 25° above the horizontal.

EXAMPLE 6.3

The ladder problem

A uniform ladder is standing on rough ground and leaning against a smooth wall at an angle of 60° to the ground. The ladder has length 4 m and mass 15 kg. Find the normal reaction forces at the wall and ground and the friction force at the ground.

Solution The diagram shows the forces acting on the ladder. The forces are in newtons.

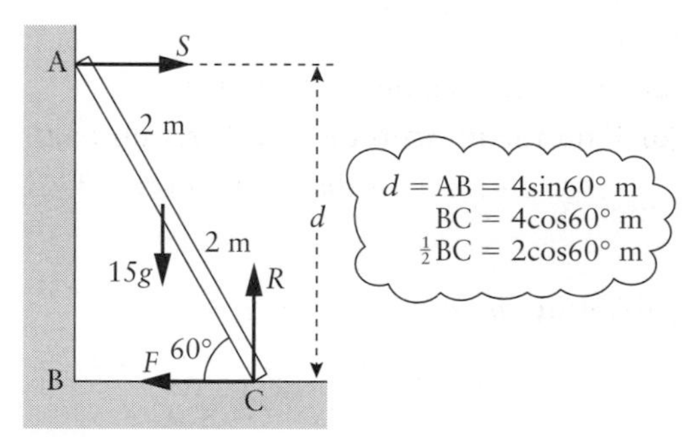

FIGURE 6.12

The diagram shows that there are three unknown forces S, R and F so we need three equations from which to find them. If the ladder remains at rest (in equilibrium) then the resultant force is zero and the resultant moment is zero. These two conditions provide the three necessary equations.

Equilibrium of horizontal components $\qquad S - F = 0 \qquad$ ①

Equilibrium of vertical components $\qquad R - 15g = 0 \qquad$ ②

Moments about the foot of the ladder

$$R \times 0 + F \times 0 + 15g \times 2\cos 60° - S \times 4\sin 60° = 0$$

$$\Rightarrow \quad 147 - 4S\sin 60° = 0 \qquad ③$$

$$\Rightarrow \quad S = \frac{147}{4\sin 60°} = 42.4$$

From ① $F = S = 42.4$
From ② $R = 147$

The force at the wall is 42.4 N, those at the ground are 42.4 N horizontally and 147 N vertically.

EXAMPLE 6.4

A ladder AB of mass m kg and length $2l$ m rests in limiting equilibrium on a rough floor against a smooth wall. The ladder makes an angle of θ with the horizontal. Find the coefficient of friction between the ladder and the floor.

Solution The diagram shows the forces.

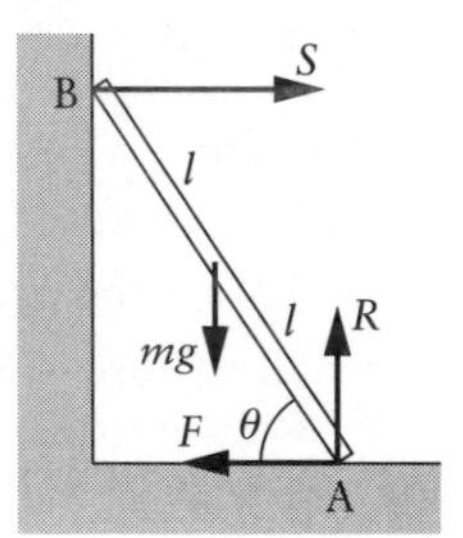

FIGURE 6.13

Resolve vertically $R = mg$ ①
Resolve horizontally $S = F$ ②
Take moments about A $mgl\cos\theta - S2l\sin\theta = 0$ ③

From ③ $S = \dfrac{mg\cos\theta}{2\sin\theta}$

Substitute in ② $F = \dfrac{mg\cos\theta}{2\sin\theta}$

Since equilibrium is limiting $F = \mu R$

so $\dfrac{mg\cos\theta}{2\sin\theta} = \mu mg$

and $\mu = \dfrac{\cos\theta}{2\sin\theta} = \dfrac{1}{2\tan\theta}$

The coefficient of friction is $\dfrac{1}{2\tan\theta}$.

EXERCISE 6A

1 Find the moment about O of each of the forces illustrated below.

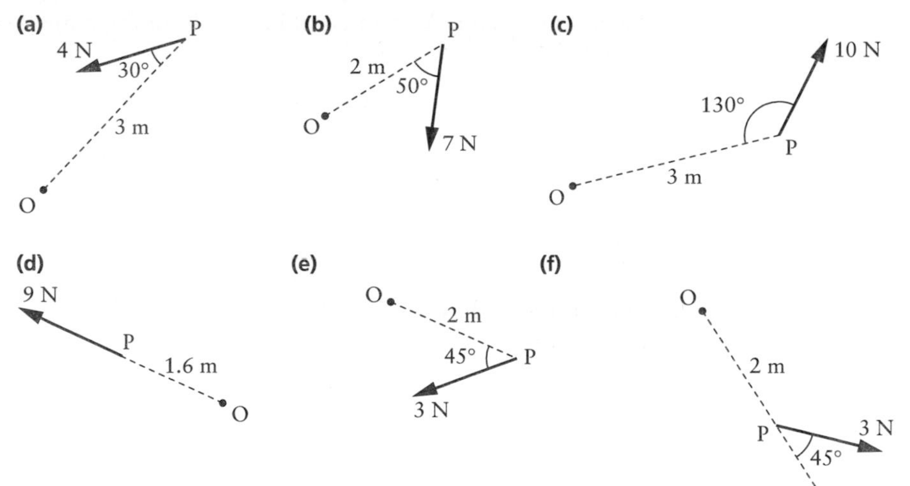

2 The diagram shows three children pushing a playground roundabout. Hannah and David want it to go one way but Rabina wants it to go the other way. Who wins?

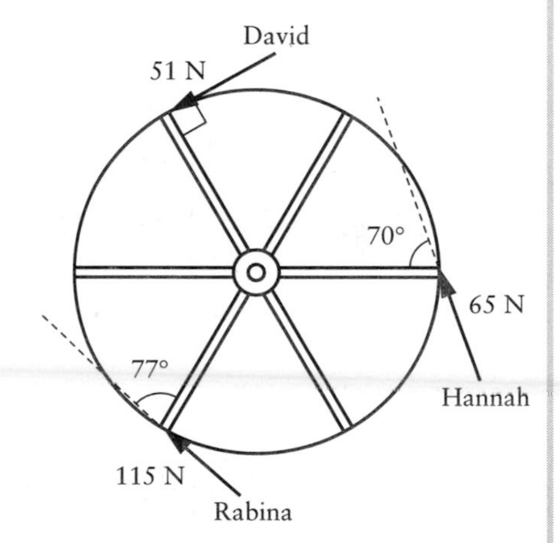

3 The operating instructions for a small crane specify that when the jib is at an angle of 25° above the horizontal, the maximum safe load for the crane is 5000 kg. Assuming that this maximum load is determined by the maximum moment that the pivot can support, what is the maximum safe load when the angle between the jib and the horizontal is:

(a) 40° **(b)** an angle θ?

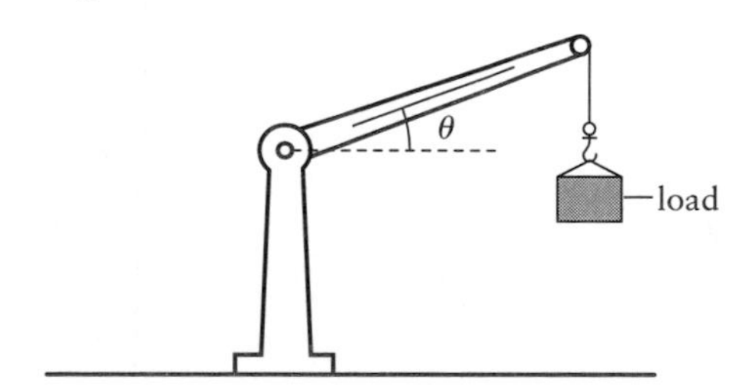

4 In each of these diagrams, a uniform beam of mass 5 kg and length 4 m, freely hinged at one end, A, is in equilibrium. Find the magnitude of the force T in each case.

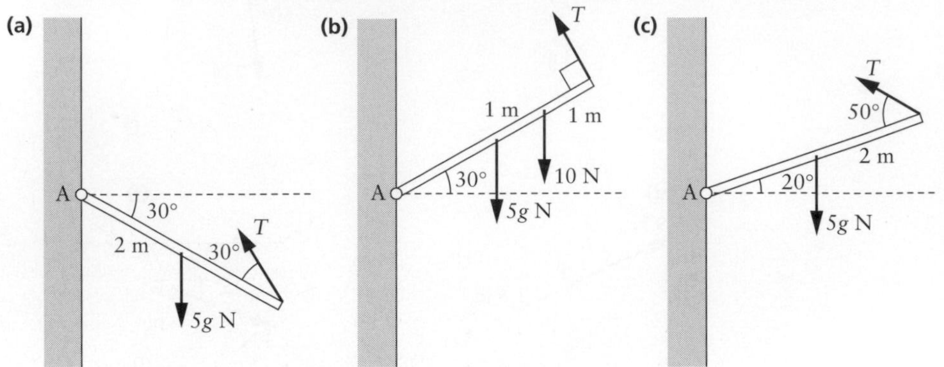

5 The diagram shows a uniform rectangular sign ABCD, 3 m × 2 m, of weight 20 N. It is freely hinged at A and supported by the string CE, which makes an angle of 30° with the horizontal. The tension in the string is T (in N).

(a) Resolve the tension T into horizontal and vertical components.

(b) Hence show that the moment of the tension in the string about A is given by

$$2T\cos 30° + 3T\sin 30°.$$

(c) Write down the moment of the sign's weight about A.

(d) Hence show that $T = 9.28$.

(e) Hence find the horizontal and vertical components of the reaction on the sign at the hinge, A.

You can also find the moment of the tension in the string about A as $d \times T$, where d is the length of AF as shown in the diagram.

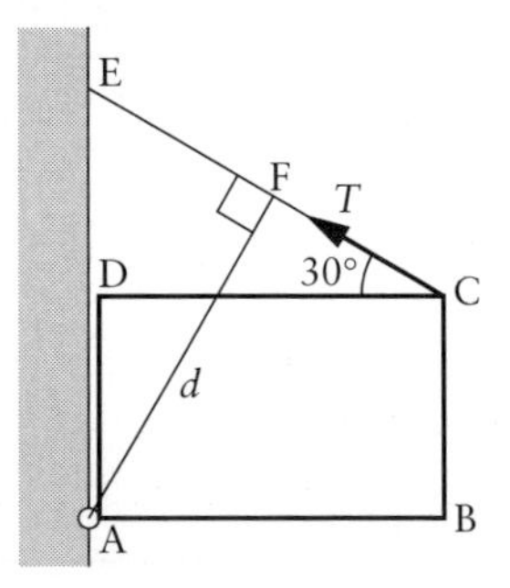

(f) Find **(i)** the angle ACD **(ii)** the length d.

(g) Show that you get the same value for T when it is calculated in this way.

6 The diagram shows a simple crane. The weight of the jib (AB) may be ignored. The crane is in equilibrium in the position shown.

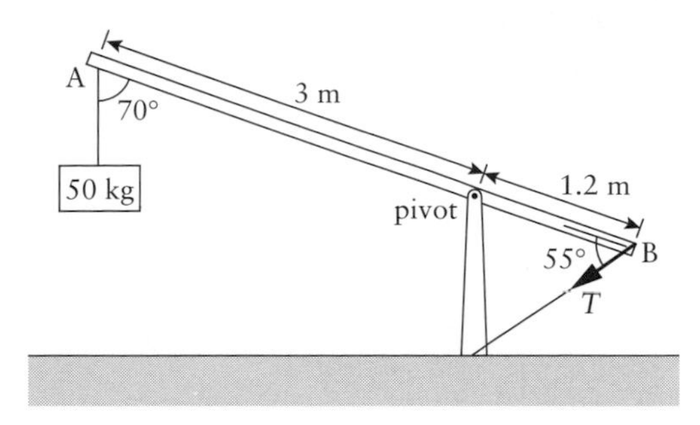

(a) By taking moments about the pivot, find the magnitude of the tension T (in N).

(b) Find the reaction of the pivot on the jib in the form of components parallel and perpendicular to the jib.

(c) Show that the total moment about the end A of the forces acting on the jib is zero.

(d) What would happen if

(i) the rope holding the 50 kg mass snapped

(ii) the rope with tension T snapped?

7 A uniform plank, AB, of mass 50 kg and length 6 m is in equilibrium leaning against a smooth wall at an angle of 60° to the horizontal. The lower end, A, is on rough horizontal ground.

(a) Draw a diagram showing all the forces acting on the plank.

(b) Write down the total moment about A of all the forces acting on the plank.

(c) Find the normal reaction of the wall on the plank at point B.

(d) Find the frictional force on the foot of the plank. What can you deduce about the coefficient of friction between the ground and the plank?

(e) Show that the total moment about B of all the forces acting on the plank is zero.

8 A uniform ladder of mass 20 kg and length $2l$ rests in equilibrium with its upper end against a smooth vertical wall and its lower end on a rough horizontal floor. The coefficient of friction between the ladder and the floor is μ. The normal reaction at the wall is S, the frictional force at the ground is F and the normal reaction at the ground is R. The ladder makes an angle α with the horizontal.

(a) Draw a diagram showing the forces acting on the ladder.

For each of the cases, **(i)** $\alpha = 60°$ **(ii)** $\alpha = 45°$

(b) find the magnitudes of S, F and R

(c) find the least possible value of μ.

9 The diagram shows a car's hand brake. The force F is exerted by the hand in operating the brake, and this creates a tension T in the brake cable. The hand brake is freely pivoted at point B and is assumed to be light.

(a) Draw a diagram showing all the forces acting on the hand brake.

(b) What is the required magnitude of force F if the tension in the brake cable is to be 1000 N?

(c) A child applies the hand brake with a force of 10 N. What is the tension in the brake cable?

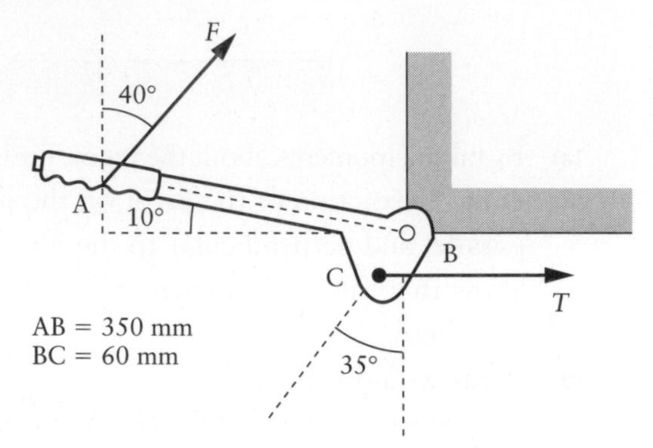

10 The diagram shows four tugs manoeuvring a ship. A and C are pushing it, B and D are pulling it.

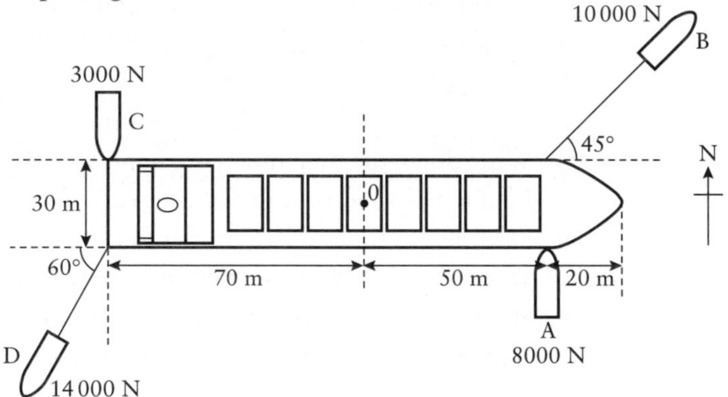

(a) Show that the resultant force on the ship is less than 100 N.

(b) Find the overall turning moment on the ship about its centre point, O.

A breeze starts to blow from the south, causing a total force of 2000 N to act uniformly along the length of the ship, at right angles to it.

(c) How (assuming B and D continue to apply the same forces) can tugs A and C counteract the sideways force on the ship by altering the forces with which they are pushing, while maintaining the same overall moment about the centre of the ship?

11 The boom of a fishing boat may be used as a simple crane. The boom AB is uniform, 8 m long and has a mass of 30 kg. It is freely hinged at the end A.

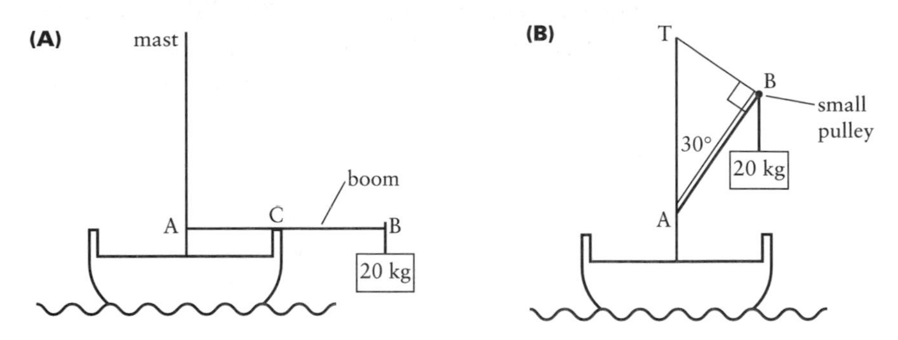

In figure (A), the boom shown is in equilibrium supported at C by the boat's rail, where the length AC is 3.5 m. The boom is horizontal and has a load of mass 20 kg suspended from the end B.

(a) Draw a diagram showing all the forces acting on the boom AB.

(b) Find the force exerted on the boom by the rail at C.

(c) Calculate the magnitude and direction of the force acting on the boom at A.

It is more usual to use the boom in a position such as the one shown in figure (B). AT is vertical and the boom is held in equilibrium by the rope section TB, which is perpendicular to it. Angle TAB = 30°. A load of mass 20 kg is supported by a rope passing over a small, smooth pulley at B. The rope then runs parallel to the boom to a fixing point at A.

(d) Find the tension in the rope section TB when the load is stationary.

[MEI, part]

12 **(a)** Draw a diagram showing the forces acting on an inclined ladder which is standing on a horizontal floor and leaning against a vertical wall. Explain why the ladder cannot be in equilibrium if the floor is frictionless, even if the wall is rough.

(b) A uniform ladder of length 8 m and mass 20 kg is inclined at 60° to the horizontal against a smooth vertical wall. A 60 kg man is standing on the ladder x m from its lower end. The horizontal floor has coefficient of friction 0.4 with the base of the ladder. The ladder is about to slip.

(i) Show that the frictional force on the ladder is $32g$ N.

(ii) Find the reaction of the wall on the ladder.

(iii) Show that x is about 6.06 m.

[MEI]

13 The diagram shows a uniform girder AB of weight 3000 N and length 6 m which has been hoisted into the air by a crane. The lengths of the ropes AC and BC are both 5 m. The tension in AC is T_1 N, and that in BC is T_2 N. The girder makes an angle α with the horizontal. The point X is directly below C and M is the mid-point of AB.

Fred, whose weight is 1000 N, was sitting on the girder when it was hoisted and now finds himself in mid-air. At the time of the question the girder and Fred are stationary. Fred is at point F where AF = 4 m and BF = 2 m.

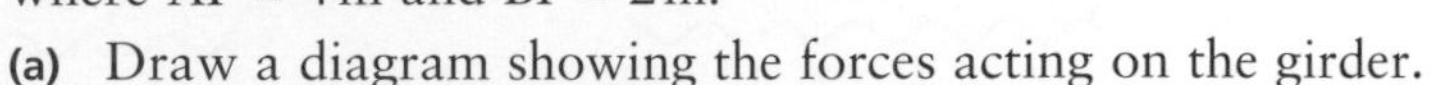

(a) Draw a diagram showing the forces acting on the girder.

(b) By taking moments about C find the distances MX and FX. Calculate the length CM and hence find α.

(c) Find the values T_1 and T_2.

(d) Fred decides to try to reach the point B. What is the value of α when he gets there?

14 The diagram shows a uniform ladder AB of mass m and length $2l$ resting in equilibrium with its upper end A against a smooth vertical wall and its lower end B on a smooth inclined plane. The inclined plane makes an angle θ with the horizontal and the ladder makes an angle ϕ with the wall.

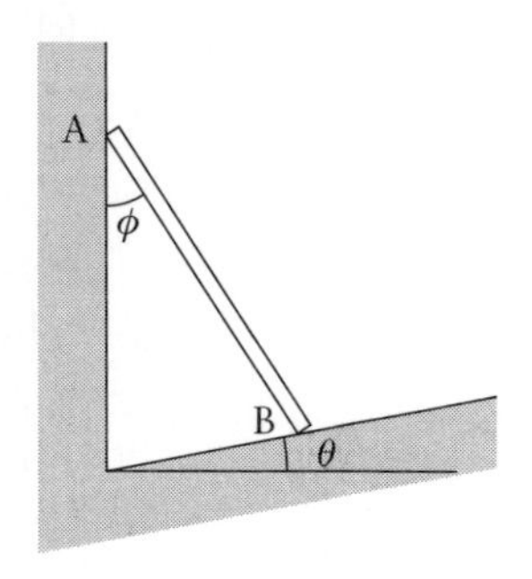

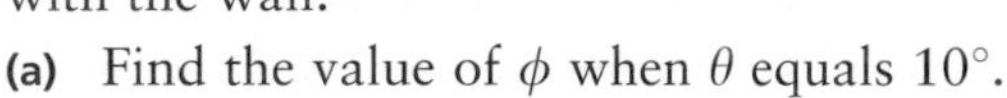

(a) Find the value of ϕ when θ equals 10°.

(b) What is the relationship between ϕ and θ?

Exercise 6B

Examination-style questions

1 The diagram shows the mast and boom of a boat. The boom may be modelled as a horizontal uniform rod of mass 30 kg and is freely hinged to the mast at A. It is supported by a light rope BC. AB = 6 m, AC = 12m and angle BAC = 90°. Find the tension in the rope.

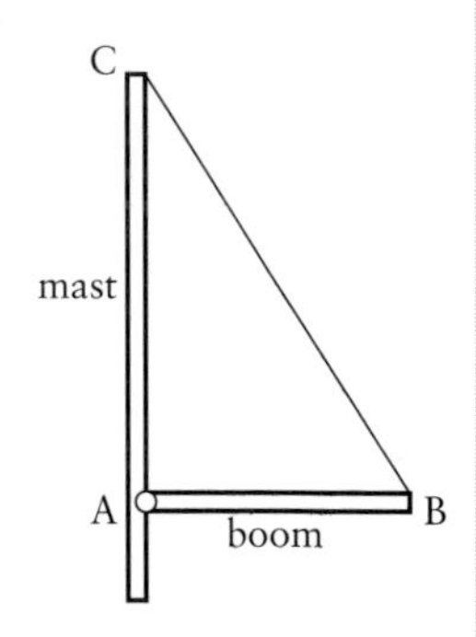

2 A mop may be modelled as a uniform rod of length 1.5 m and mass 0.5 kg with a particle of mass 1 kg attached at one end (the head). The mop is standing in limiting equilibrium on a rough horizontal floor with its head leaning against a smooth vertical wall. The coefficient of friction between the floor and the mop is 0.3. Find the angle between the mop and the horizontal.

3 A uniform rod AB of length 2 m and mass 20 kg is smoothly hinged to a vertical wall at A and supported at an angle of 40° to the downward vertical by a light string BC. The point C is 2 m directly above A on the wall.

(a) Find the tension in the string BC.

(b) Find the magnitude of the force exerted on the rod by the hinge.

4 A uniform rod AB of mass 10 kg is smoothly hinged to a vertical wall at A and supported at an angle of θ to the upward vertical by a light string BC, where $\sin\theta = \frac{2}{7}$. C is vertically above A on the wall and the angle ABC is 90°.

(a) Find the tension in the string BC.

(b) Find the magnitude of the force exerted on the rod by the hinge.

5 A uniform rod AB of mass 8 kg and length 1.5 m is smoothly hinged to a vertical wall at A. It is supported in a horizontal position by a light string CD, where C is a point on the wall 0.7 m vertically above A and D is a point on the rod 1.2 m from A. The rod carries a load of 15 kg at B.

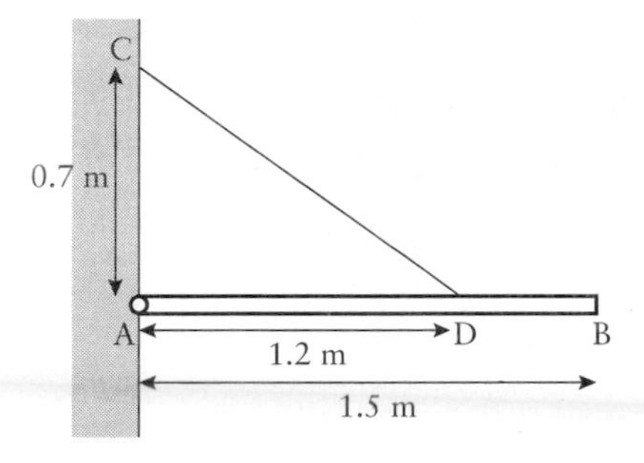

(a) Find the tension in the string CD.

(b) Find the vertical component of the force exerted on the rod by the hinge at A, stating whether it is up or down.

(c) Show that your answer to part **(b)** is independent of the distance AC.

6 While a trestle table is being erected, the top is resting on one trestle only, with one end resting on rough horizontal ground. The top is 4 m long, and may be modelled as a uniform plank of mass 10 kg. The trestle is 0.7 m high and the point of contact between the trestle and the top is 3 m along the top. The contact between the top and the trestle may be treated as smooth.

(a) Find the angle between the top and the ground.

(b) Show that the normal contact force between the top and the trestle is 63.5 N, and find the normal contact force and the friction force where the top meets the ground.

(c) Given that the top is in limiting equilibrium, find the coefficient of friction between the top and the ground.

7 A uniform ladder of mass 20 kg and length 5 m is standing on rough horizontal ground and leaning against the top corner of a smooth vertical wall. The foot of the ladder is 2 m from the foot of the wall and the point of contact between the ladder and the wall is 4 m from the foot of the ladder.

(a) Find the reaction between the wall and the ladder.

(b) Find the normal reaction of the ground on the ladder.

(c) Find the friction between the ground and the ladder.

(d) What can you say about the coefficient of friction between the ground and the ladder?

8 A uniform ladder of mass M and length 6 m rests on rough horizontal ground and leans against a smooth vertical wall. The coefficient of friction between the ground and the ladder is 0.3.

(a) Find the minimum angle which the ladder can make with the horizontal and remain in equilibrium.

The ladder is now moved so that the angle between the ladder and the ground is $65°$.

(b) Find, in terms of M, the mass of the heaviest person who can safely climb to the top of the ladder.

9 A uniform ladder of mass m and length $2l$ is standing on rough horizontal ground and leaning against a rough vertical wall. The coefficient of friction between the ladder and the ground is 0.3 and that between the ladder and the wall is 0.2. The ladder is in limiting equilibrium.

(a) Show that the normal reaction between the ladder and the wall is

$$\frac{15mg}{53}.$$

(b) Find the angle between the ladder and the ground.

10 A uniform ladder of mass 10 kg is standing on rough horizontal ground and leaning against a cuboidal box of mass m which in turn leans against a rough vertical wall, but does not reach the ground. The contact between the ladder and the box is smooth, but the coefficient of friction between the box and the wall is μ, the same as the coefficient of friction between the ladder and the ground. The ladder is in limiting equilibrium and makes an angle of $70°$ with the ground.

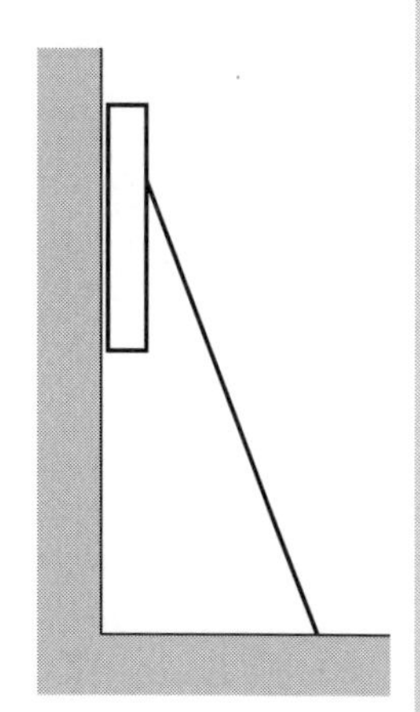

(a) Find μ.

(b) Find the maximum value of m if the box is to remain in equilibrium.

11 A uniform ladder AB of mass m and length $2a$, has one end A on rough horizontal ground. The other end B rests against a smooth vertical wall. The ladder is in a vertical plane perpendicular to the wall. The ladder makes an angle α with the horizontal, where $\tan\alpha = \frac{4}{3}$. A child of mass $2m$ stands on the ladder at C where AC $= \frac{1}{2}a$, as shown in the diagram. The ladder and the child are in equilibrium.

By modelling the ladder as a rod and the child as a particle, calculate the least possible value of the coefficient of friction between the ladder and the ground.

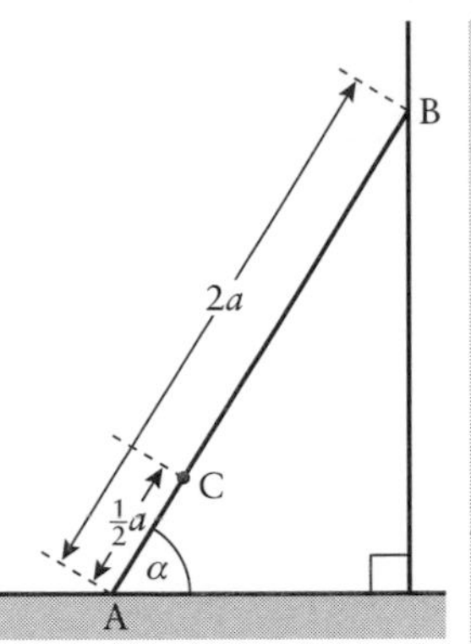

[Edexcel]

12

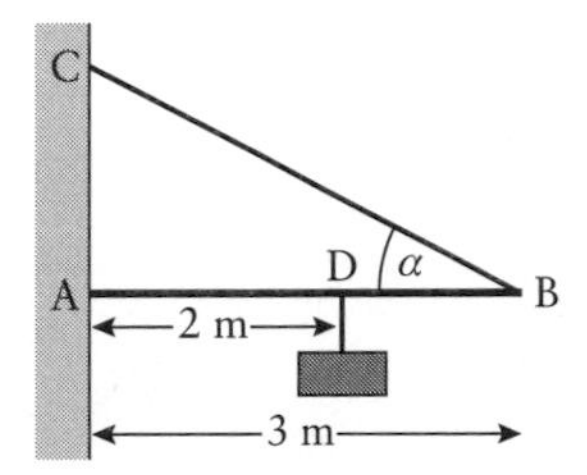

A uniform steel girder AB, of mass 40 kg and length 3 m, is freely hinged at A to a vertical wall. The girder is supported in a horizontal position by a steel cable attached to the girder at B. The other end of the cable is attached to point C vertically above A on the wall, with $\angle ABC = \alpha$, where $\tan\alpha = \frac{3}{4}$. A load of mass 60 kg is suspended by another cable from the girder at the point D, where AD = 2 m, as shown in the diagram. The girder remains horizontal and in equilibrium. The girder is modelled as a rod, and the cables as light inextensible strings.

(a) Show that the tension in the cable BC is 980 N.

(b) Find the magnitude of the reaction on the girder at A.

(c) Explain how you have used the modelling assumption that the cable at D is light.

[Edexcel]

13

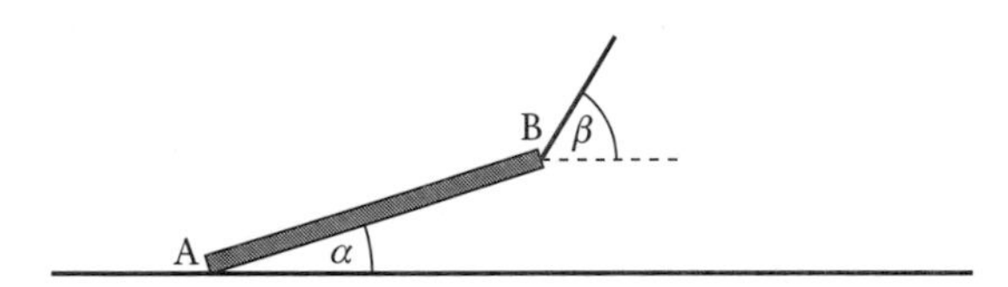

A straight log AB has weight W and length $2a$. A cable is attached to one end B of the log. The cable lifts the end B off the ground. The end A remains in contact with the ground, which is rough and horizontal. The log is in limiting equilibrium. The log makes an angle α to the horizontal, where $\tan\alpha = \frac{5}{12}$.

The cable makes an angle β to the horizontal, as shown in the diagram. The coefficient of friction between the log and the ground is 0.6. The log is modelled as a uniform rod and the cable as light.

(a) Show that the normal reaction on the log at A is $\frac{2}{5}W$.

(b) Find the value of β.

The tension in the cable is kW.

(c) Find the value of k.

[Edexcel]

KEY POINTS

1 The moment of a force F about a point O is given by the product Fd where d is the perpendicular distance from O to the line of action of the force.

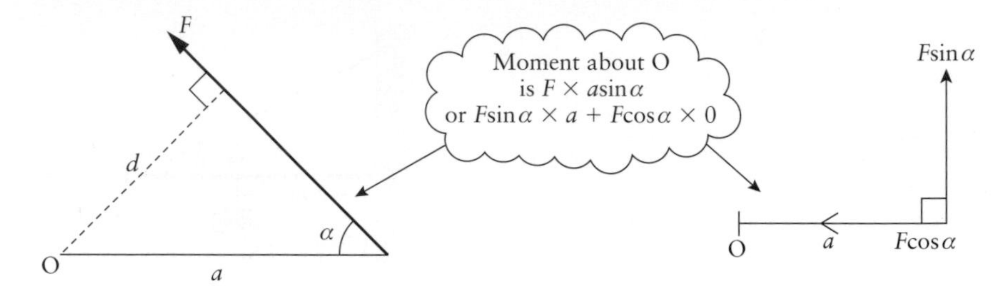

2 The S.I. unit for moment is the newton metre (Nm).

3 Anticlockwise moments are usually called positive, clockwise negative.

4 If a body is in equilibrium the sum of the moments of the forces acting on it, about any point, is zero.

5 When three non-parallel forces are in equilibrium, their lines of action are concurrent.

ANSWERS

CHAPTER 1

EXERCISE 1A (Page 4)

1 (a) (i)

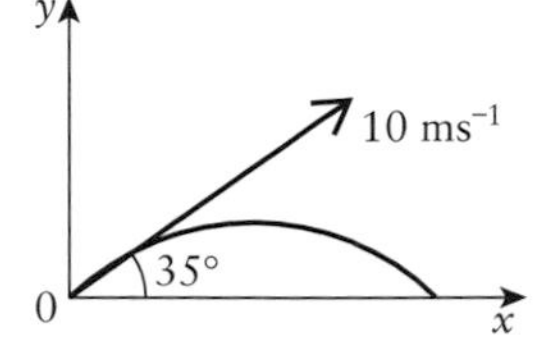

(ii) $u_x = 8.2$

$u_y = 5.7$

(iii) $v_x = 8.2$

$v_y = 5.7 - 9.8t$

(iv) $x = 8.2t$

$y = 5.7t - 4.9t^2$

(b) (i)

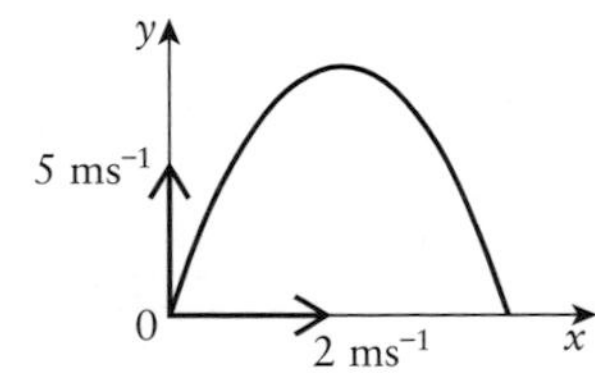

(ii) $u_x = 2$

$u_y = 5$

(iii) $v_x = 2$

$v_y = 5 - 9.8t$

(iv) $x = 2t$

$y = 5t - 4.9t^2$

(c) (i)

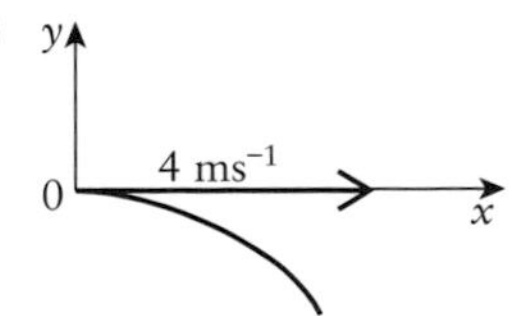

(ii) $u_x = 4$

$u_y = 0$

(iii) $v_x = 4$

$v_y = -9.8t$

(iv) $x = 4t$

$y = -4.9t^2$

(d) (i)

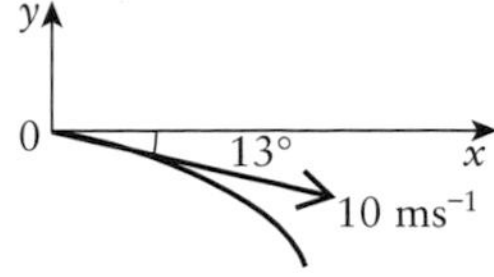

(ii) $u_x = 9.7$

$u_y = -2.2$

(iii) $v_x = 9.7$

$v_y = -2.2 - 9.8t$

(iv) $x = 9.7t$

$y = -2.2t - 4.9t^2$

(e) (i)

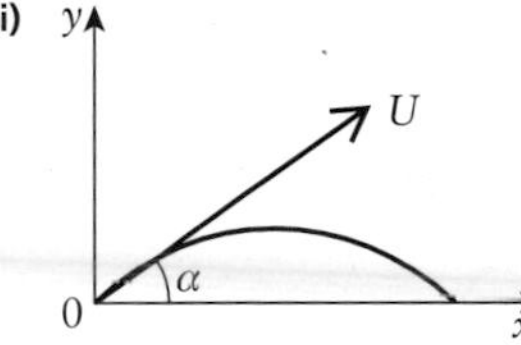

(ii) $u_x = U\cos\alpha$

$u_y = U\sin\alpha$

(iii) $v_x = U\cos\alpha$

$v_y = U\sin\alpha - gt$

(iv) $x = Ut\cos\alpha$

$y = Ut\sin\alpha - \frac{1}{2}gt^2$

(f) (i)

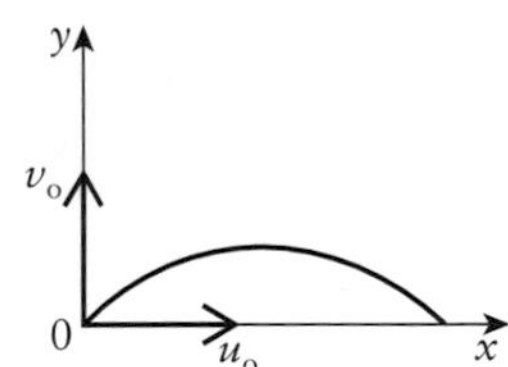

(ii) $u_x = u_o$

$u_y = v_o$

(iii) $v_x = u_o$

$v_y = v_o - gt$

(iv) $x = u_o t$

$y = v_o t - \frac{1}{2}gt^2$

2 (a) (i) 1.5 s

(ii) 11 m

(b) (i) 0.5 s

(ii) 1.25 m

3 (a) (i) 4 s
(ii) 80 m
(b) (i) 0.88 s
(ii) 2.21 m

Exercise 1B (Page 6)

1 (a) (i)

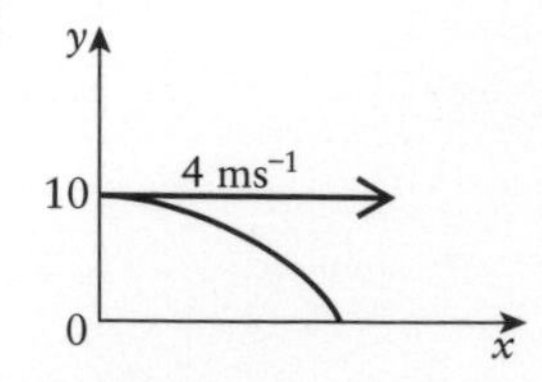

(ii) 4, $-9.8t$
(iii) $4t$, $10 - 4.9t^2$
(b) (i)

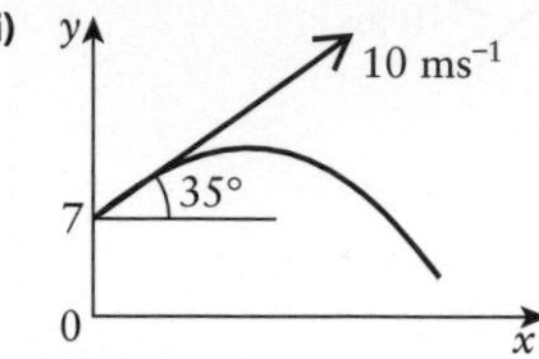

(ii) 8.2, $5.7 - 9.8t$
(iii) $8.2t$, $7 + 5.7t - 4.9t^2$
(c) (i)

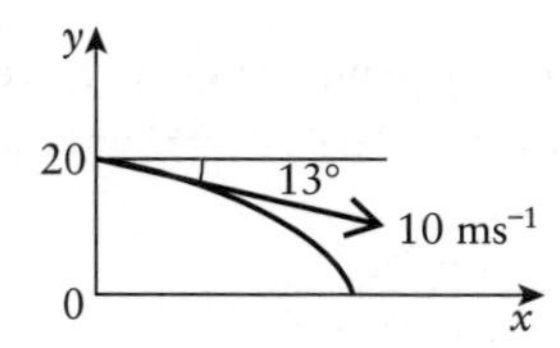

(ii) 9.7, $-2.2 - 9.8t$
(iii) $9.7t$, $20 - 2.2t - 4.9t^2$
(d) (i)

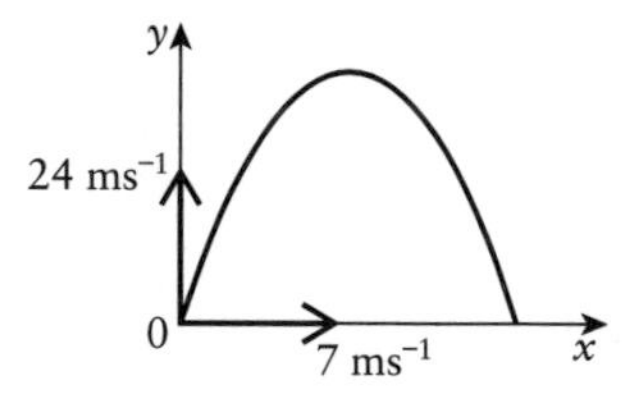

(ii) 7, $24 - 9.8t$
(iii) $7t$, $24t - 4.9t^2$
(e) (i)

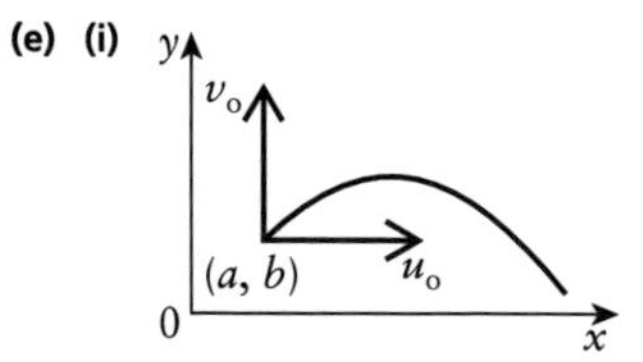

(ii) u_o, $v_o - gt$
(iii) $a + u_ot$, $b + v_ot - \frac{1}{2}gt^2$

2 (a) (i) 1.5 s
(ii) 26 m
(b) (i) 0.31 s
(ii) 10.46 m

3 (a) 2.86 m
(b) 2.86 m
(c)

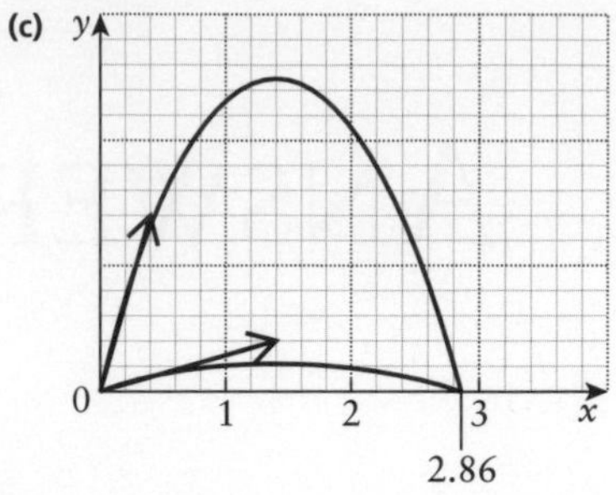

Exercise 1C (Page 12)

1 (a) 17.3, 10 ms^{-1}
(b) 0, -10 ms^{-2}
(c) 34.6 m
(d) 1 s
(e) 5 m

2 (a) 41, 28.7 ms^{-1}
(b)

t	0	1	2	3	4	5	6
x	0	41	82	123	164	205	246
y	0	24	38	42	36	21	−4.3

(c)

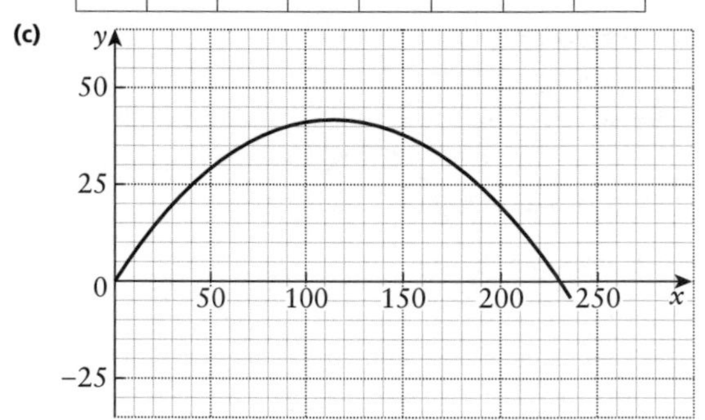

(d) 42 m, 239.7 m
(e) Air resistance and spin ignored, ball treated as a particle.

3 (a) 17.2, 8 ms^{-1}
(b) 1.64 s
(c) 28.2 m
(d) 0.82 s
(e) 3.3 m
(f) 2.72 m, no

4 (a) 10.3, 14.7 ms^{-1}
(b) 2.91 s
(c) Into the goal
(d) No

5 (a) 45.2 s
(b) 13.6 km

(c) $535\,ms^{-1}$
(d) $56°$
6 (a) $0.47\,s$
(b) $0.64\,s$
(c) $25.4\,ms^{-1}$
(d) $28.8\,ms^{-1}$
7 (a) Yes, the range is 70.4 m.
(b) $32.7\,ms^{-1}$
8 (a) $2.02\,s$
(b) No, height = 0.2 m
(c) $21.57\,ms^{-1}$
(d) Spin causes the ball to rise more.
9 (a) (i) 34.6 m
(ii) 39.4 m
(iii) 40 m
(iv) 39.4 m
(v) 34.6 m
(b) $80\sin\alpha\cos\alpha = 80\cos(90 - \alpha)\sin(90 - \alpha)$
(c) 57.9
10 (a) $26.1\,ms^{-1}$
(b) $27.35\,ms^{-1}$
(c) $26.15 < u < 26.88$
11 25.5 m

EXERCISE 1D (Page 15)

1 (a) $y = \frac{5}{16}x^2$
(b) $y = 6 + 0.4x - 0.2x^2$
(c) $y = -14 + 17x - 5x^2$
(d) $y = 5.8 + 2.4x - 0.2x^2$
(e) $y = 2x - \dfrac{gx^2}{2u^2}$
2 (a) $x = 40t$
(c) (d)

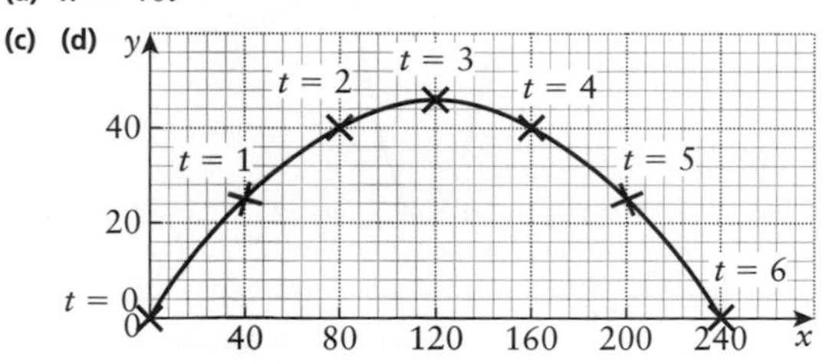

3 (a) $y = \frac{3}{4}x - \frac{1}{320}x^2$
(b) Air resistance would reduce x.

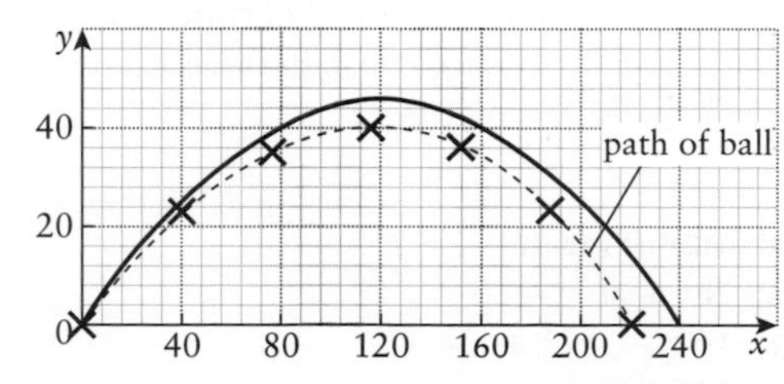

(c) Yes, horizontal acceleration = $-\,0.5\,ms^{-2}$

EXERCISE 1E (Page 19)

1 (a) $(21.2t,\ 21.2t - 5t^2)$
(c) 8.9
(d) 29.7 or 61.2
2 (b) 6.9 m
3 (b) 24.1 m
(c) Yes, $y = 2.4$ m
4 (a) $y = 1 + 0.7x - \frac{7.45x^2}{u^2}$
(b) $u > 7.8\,ms^{-1}$
(c) $u < 8.5\,ms^{-1}$
5 (b) $63°, 83°$
(c) (i) Yes
(ii) No
6 (a) $y = 2 + x\tan\alpha - \frac{x^2}{320}(1 + \tan^2\alpha)$
(b) $87°, -3.1°$
(c) (i) No
(ii) Yes
7 (a) $x = 25t\cos\alpha,\ y = 1 + 25t\sin\alpha - 5t^2$
(b) $y = 1 + x\tan\alpha - \frac{x^2}{125}(1 + \tan^2\alpha)$
(c) (i) No
(ii) Yes
(d) $13°, 83°$
8 (a) $x = 15t\cos\alpha$
(d) No, 45° gives the maximum range level with the point of projection. In this case 44° gives greater X.
9 (b) 1.8, 0.6
(c) 22.5 m when $\tan\alpha = 1.2$
(d) 15.4 m (taking the high route)
10 (a) $y = \frac{4x}{3} - \frac{x^2}{180}$
(b) $\frac{y}{x} = \tan\beta = \frac{1}{2}$
(c) (150, 75)
(d) 170 m
(e) 140 m
(f) No

EXERCISE 1F (page 24)

1 (a) $0.404\,s$
(b) $3.71\,ms^{-1}$
(c) $28.1°$
2 (a) 221 m
(b) 31.9 m
(c) Yes
3 (a) $20.9\,ms^{-1}$
(b) $25.6\,ms^{-1}$
(c) Include air resistance; include effect of spin; allow for diagonal serve.
4 (a) $10.9\,s$
(b) 393 m
(c) $69\,ms^{-1}$
(d) $58.6°$

5 (a) 22.5°
(b) 1.56 s
(c) 28.9 m
(d) Yes

6 (a) $6.71\ \text{ms}^{-1}$
(b) 0.685 s
(c) $16.8\ \text{ms}^{-1}$
(d) 21.8°

7 (a) 2.14 m
(b) 9.15ms^{-1}
(c) 64.1°

8 (a) $6.57\ \text{ms}^{-1}$
(b) $7\ \text{ms}^{-1}$
(c) 0.467 s and 0.962 s
(d) $1.04\ \text{ms}^{-1} \leqslant v \leqslant 2.14\ \text{ms}^{-1}$

9 (a) $0.225\ \text{s} \leqslant t \leqslant 1.82\ \text{s}$
(b) Under 5.39 m or between 43.6 m and 50.2 m.

10 (b) 0.877°
(c) 1.23 m

11 (a) 20 s
(b) 33.2°
(c) 18.1°

12 (a) 2 s
(b) 31.5 m

13 (b) 4 m

14 (a) $2\frac{6}{7}$ s
(b) 1.52 m
(c) Include air resistance.

CHAPTER 2

EXERCISE 2A (Page 32)

1 (a) (i) $v = 2 - 2t$
(ii) 10, 2
(iii) 1, 11
(b) (i) $v = 2t - 4$
(ii) 0, −4
(iii) 2, −4
(c) (i) $3t^2 - 10t$
(ii) 4, 0
(iii) 0, 4 and $3\frac{1}{3}$, −14.5

2 (a) (i) 4
(ii) 3, 4
(b) (i) $12t - 2$
(ii) 1, −2
(c) (i) 7
(ii) −5, 7

3 $v = 4 + t, a = 1$

4 (a) (i) $v = 15 - 10t, a = -10$
(ii)

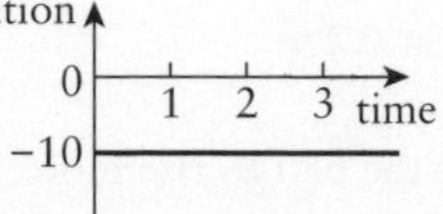

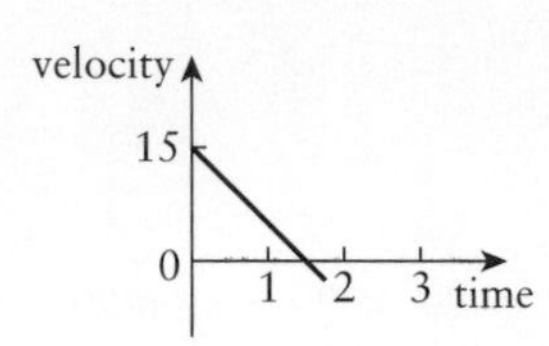

(iii) The acceleration is the gradient of the velocity–time graph.
(iv) The acceleration is constant; the velocity decreases at a constant rate.

(b) (i) $v = 18t^2 - 36t - 6, a = 36t - 36$
(ii)

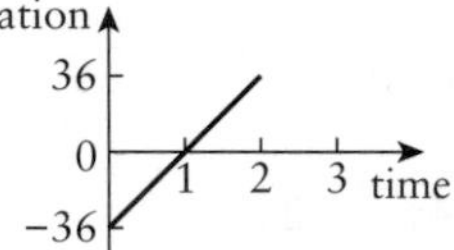

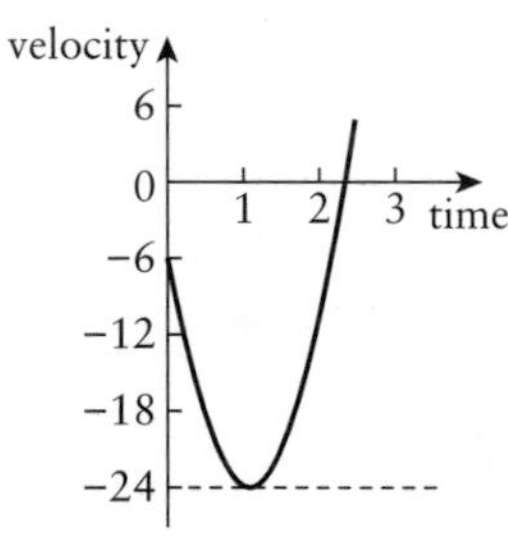

(iii) The acceleration is the gradient of the velocity–time graph; velocity is at a minimum when the acceleration is 0.
(iv) It starts in the negative direction. v is initially −6 and decreases to −24 before increasing rapidly to zero, where the object turns to move in the positive direction.

EXERCISE 2B (Page 36)

1 (a) $2t^2 + 3t$
(b) $1.5t^4 - \frac{2}{3}t^3 + t + 1$
(c) $\frac{7}{3}t^3 - 5t + 2$

2 **(a)**

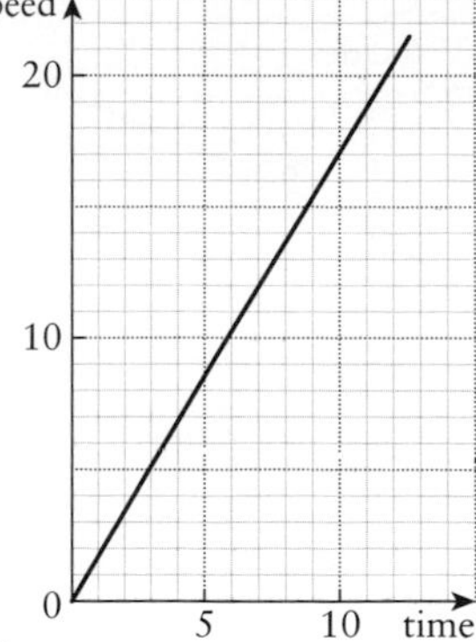

(b) 85 m

3 **(a)** When $t = 6$

(b) 972 m

4 **(a)** 4.47 s

(b) 119 m

5 **(a)** $v = 10t + \frac{3}{2}t^2 - \frac{1}{3}t^3,\ x = 5t^2 + \frac{1}{2}t^3 - \frac{1}{12}t^4$

(b) $v = 2 + 2t^2 - \frac{2}{3}t^3,\ x = 1 + 2t + \frac{2}{3}t^3 - \frac{1}{6}t^4$

(c) $v = -12 + 10t - 3t^2,\ x = 8 - 12t + 5t^2 - t^3$

6 **(a)** $15 - 10t$

(b) 11.5 m, $+5\text{ ms}^{-1}$, 5 ms^{-1}; 11.5 m, -5 ms^{-1}, 5 ms^{-1}

(c)

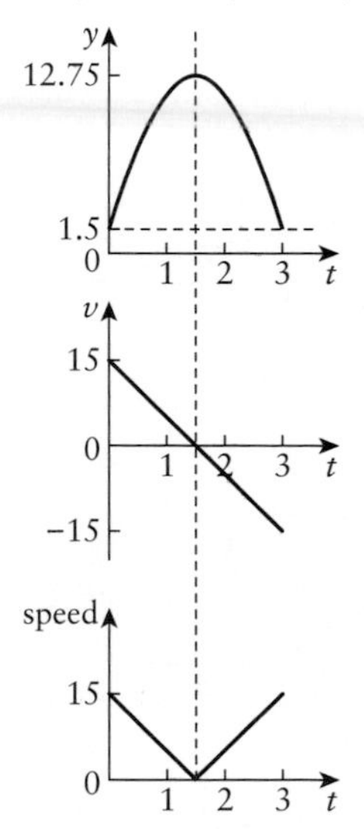

(d) 3 s

(e) The expression does not equal the distance travelled because of changes in direction. The expression gives the displacement from the original position, which equals 0.

7 **(a)** -3 m, -1 ms^{-1}, 1 ms^{-1}

(b) **(i)** 1 s

(ii) 2.15 s

(c)

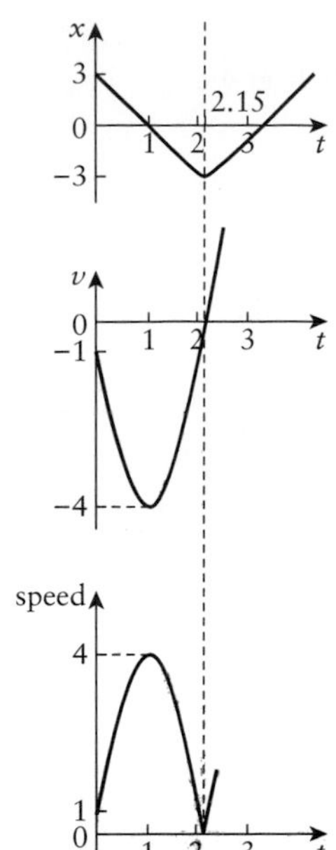

(d) The object moves in a negative direction from 3 m to -3 m then moves in a positive direction with increasing speed.

8 2 s

9 **(a)** $v = 4 + 4t - t^2,\ x = 4t + 2t^2 - \frac{1}{3}t^3$

(b)

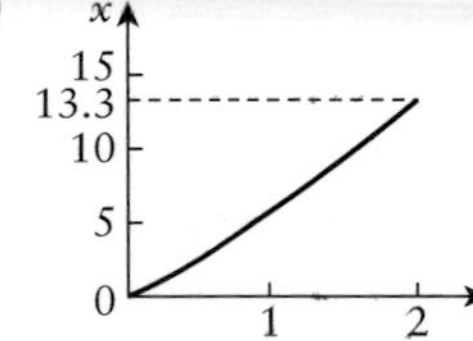

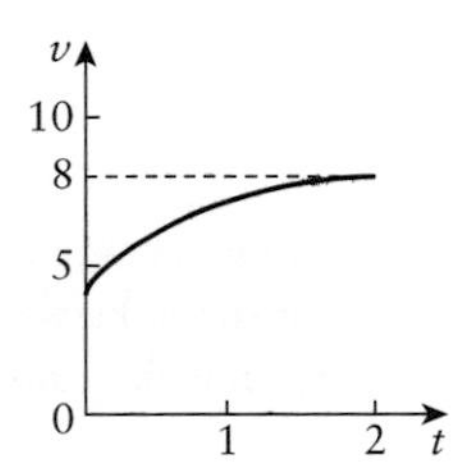

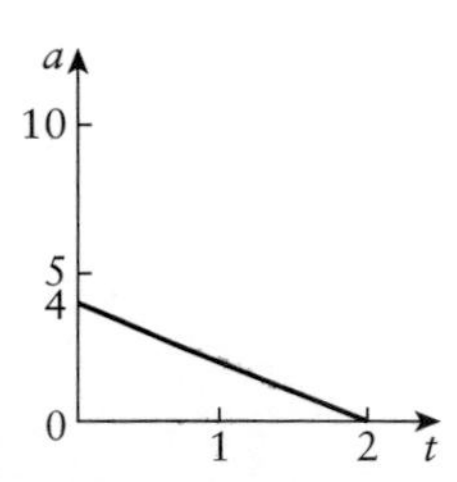

(c) The object starts at origin and moves in a positive direction with increasing speed reaching a maximum speed of 8 ms^{-1} after 2 s.

10 (a) 0, 10.5, 18, 22.5, 24

(b) The ball reaches the hole at 4 s.

(c) $-3t + 12\ \mathrm{ms^{-1}}$

(d) $0\ \mathrm{ms^{-1}}$

(e) $-3\ \mathrm{ms^{-2}}$

11 (a) Andrew: $10\ \mathrm{ms^{-1}}$, Elizabeth: $9.6\ \mathrm{ms^{-1}}$

(b)

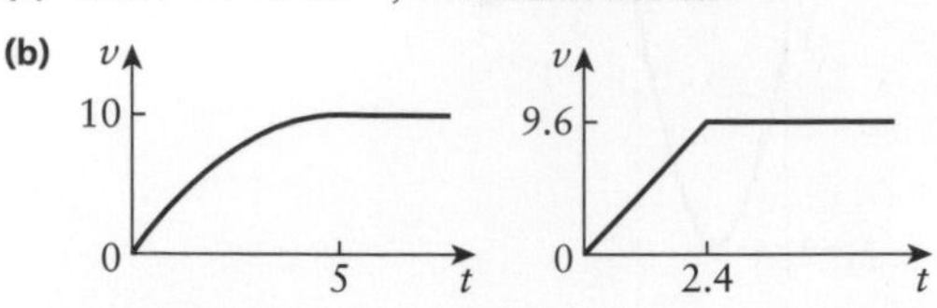

(c) 11.52 m

(d) 11.62 s

(e) Elizabeth by 0.05 s and 0.5 m

(f) Andrew wins.

12 (a)

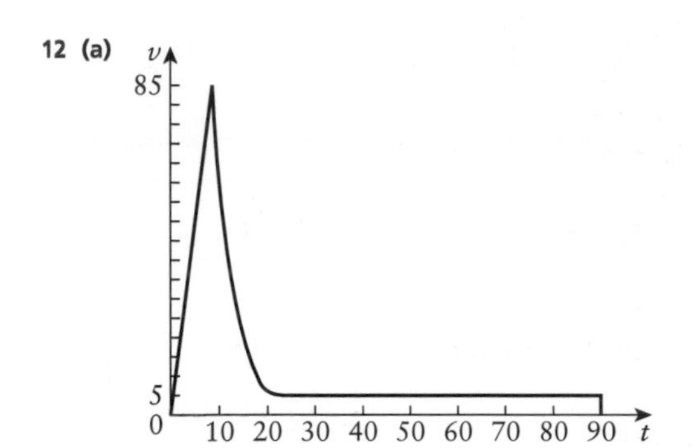

Christine is in free fall until $t = 10$ s then the parachute opens and she slows down to her terminal velocity of $5\ \mathrm{ms^{-1}}$.

(b) 1092 m

(c) $8.5\ \mathrm{ms^{-2}}$, $1.6t - 32$, $0\ \mathrm{ms^{-2}}$, $16\ \mathrm{ms^{-2}}$

13 (a) Between P and Q the train speeds up with gradually decreasing acceleration. Between Q and R it is travelling at a constant speed. Between S and T the train is slowing down with a constant deceleration.

(b) −0.000025, 0.05

(c) $50\ \mathrm{ms^{-1}}$

(d) $0\ \mathrm{ms^{-1}}$

(e) $111\frac{2}{3}$ km

EXERCISE 2C (Page 45)

1 (a) $4t\mathbf{i} + 8\mathbf{j}$

(b) (0, 0), (2, 8), (8, 16), (18, 24), (32, 32)

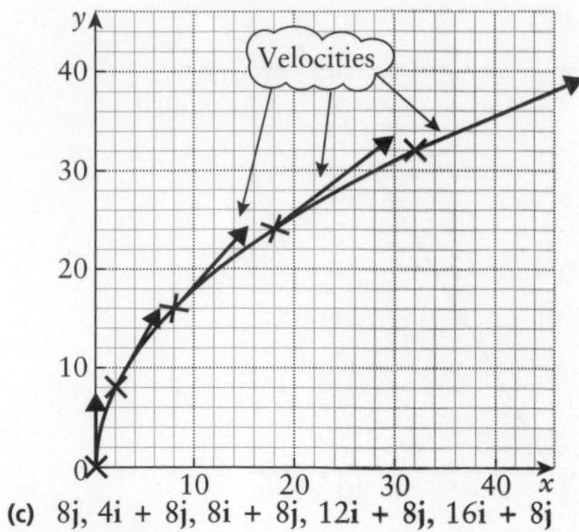

(c) $8\mathbf{j}$, $4\mathbf{i} + 8\mathbf{j}$, $8\mathbf{i} + 8\mathbf{j}$, $12\mathbf{i} + 8\mathbf{j}$, $16\mathbf{i} + 8\mathbf{j}$

(d) $21.5\ \mathrm{ms^{-1}}$

2 $\mathbf{v} = -4\mathbf{i} - 5\mathbf{j}$, $\mathbf{a} = 0$

3 $4.47\ \mathrm{ms^{-2}}$, $-153°$

4 (a) $\frac{1}{20}t^2\mathbf{i} + \frac{1}{30}t^3\mathbf{j}$

(b) $5\mathbf{i} + 33.3\mathbf{j}$ m

5 $\mathbf{v} = 2t^2\mathbf{i} + (6t - t^2)\mathbf{j}$, $\mathbf{r} = \frac{2}{3}t^3\mathbf{i} + (3t^2 - \frac{1}{3}t^2)\mathbf{j}$

6 $8.11\ \mathrm{ms^{-1}}$

7 (a) Initial velocity $= 3.54\mathbf{i} - 3.54\mathbf{j}$

(b) $\mathbf{v} = 8.54\mathbf{i} + 11.46\mathbf{j}$; $\mathbf{r} = 52.0\mathbf{i} + 14.6\mathbf{j}$ m

8 (a) $\mathbf{v} = 15\mathbf{i} + (16 - 10t)\mathbf{j}$; $\mathbf{a} = -10\mathbf{j}$

(b) 1.6 s

(c) $22.8\ \mathrm{ms^{-1}}$

9 (a) (0, 2), (4, 2), (8, 1.6), (12, 0.8), (16, −0.4)

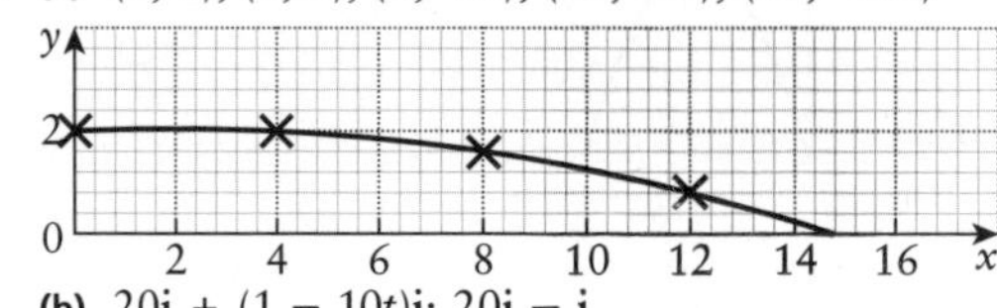

(b) $20\mathbf{i} + (1 - 10t)\mathbf{j}$; $20\mathbf{i} - \mathbf{j}$

(c) $-10\mathbf{j}$

10 (a)

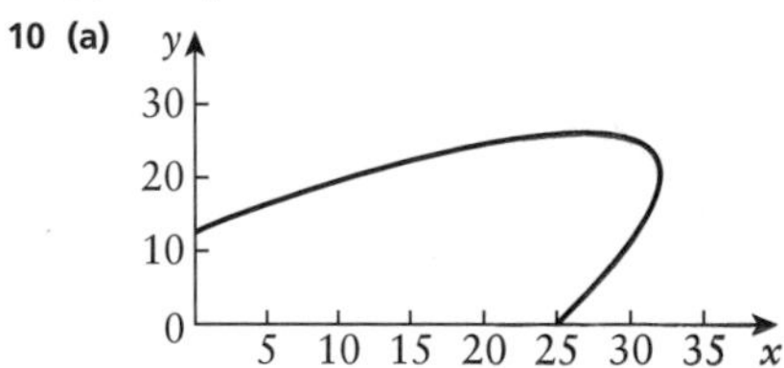

(b) 5 s

(c) $33.5\ \mathrm{ms^{-1}}$, $63.4°$

11 (a) A: $v\sin 35°\,\mathbf{i} + v\cos 35°\,\mathbf{j}$; B: $-8.66\mathbf{i} + 5\mathbf{j}$

(b) A: $vt\sin 35°\,\mathbf{i} + vt\cos 35°\,\mathbf{j}$;
B: $(-8.66t + 5)\mathbf{i} + 5t\mathbf{j}$

(c) $6.1\ \mathrm{km\,h^{-1}}$

(d) 24.7 minutes

12 (a) $2t^2\mathbf{i} + 4t\mathbf{j}$

(b) $\frac{2}{3}t^3\mathbf{i} + 2t^2\mathbf{j}$

(c) $\frac{1}{6}t^4\mathbf{i} + \frac{2}{3}t^3\mathbf{j}$

(d) $\mathbf{F} = 4\mathbf{i} + 4\mathbf{j}$, $\mathbf{a} = 8\mathbf{i} + 8\mathbf{j}$, $\mathbf{v} = \frac{16}{3}\mathbf{i} + 8\mathbf{j}$,
$\mathbf{r} = \frac{8}{3}\mathbf{i} + \frac{16}{3}\mathbf{j}$, speed $9.6\ \mathrm{ms^{-1}}$, 6.0 m from O

13 (a)

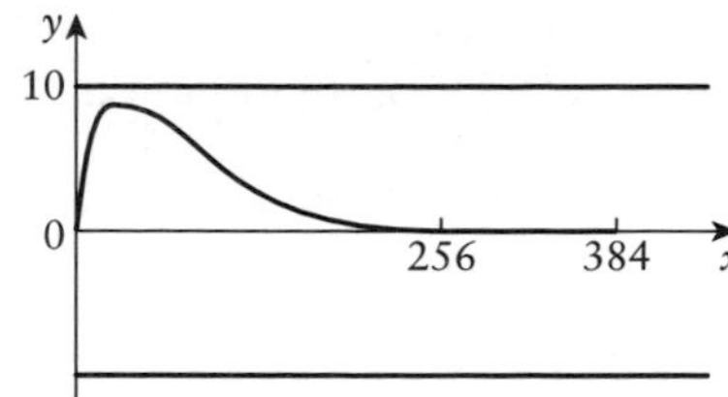

No, maximum $y = 9.48$

(b) $\mathbf{v} = 8t\mathbf{i} + \frac{1}{8}(64 - 32t + 3t^2)\mathbf{j}$;
$\mathbf{a} = 8\mathbf{i} + \frac{1}{8}(6t - 32)\mathbf{j}$ for $0 \leqslant t \leqslant 8$;
$\mathbf{v} = 64\mathbf{i}$; $\mathbf{a} = 0$ for $8 < t \leqslant 20$

(c) $1200\mathbf{i} + (112.5t - 600)\mathbf{j}$ N

(d) To overcome air resistance.

EXERCISE 2D (Page 48)

1 (a) $t^2 - 5t + 4\ \text{ms}^{-1}$
(b) $4\frac{5}{6}$ m
2 (a) 0 s, 4 s
(b) $2\frac{2}{3}$ m
(c) $2\ \text{ms}^{-2}$
3 (a) $\frac{4}{27}$ m, 0 m
(b) $\frac{2}{3}$ s
(c) $2\frac{8}{27}$ m
4 (a) $v = \dfrac{3t^2}{2} - \dfrac{t^3}{4}$, $x = \dfrac{t^3}{2} - \dfrac{t^4}{16}$
(b) $3\ \text{ms}^{-2}$
(c) $8\ \text{ms}^{-1}$
(d) 16 m
5 (a) $11\frac{1}{3}$ s

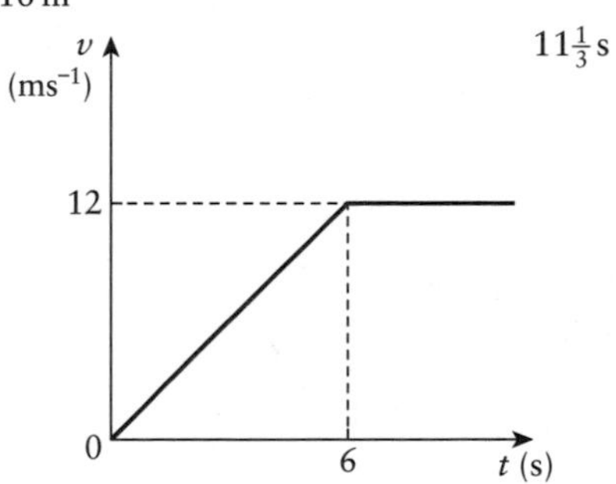

(b) $v = 3.5t - 0.25t^2$, $x = \dfrac{7t^2}{4} - \dfrac{t^3}{12}$
(c) 100.8 m. The race has finished.
(d) $6 < t < 8$
6 (a) $(3t^2 - 4)\mathbf{i} + (3t + 1)\mathbf{j}$
(b) 1.5 s
7 (a) $2\mathbf{i} + (3 - 2t)\mathbf{j}$
(b) 153.4°
8 (a) $(-2t + 4)\mathbf{i} + (\frac{1}{2}t^2 - t + 1)\mathbf{j}$
(b) $(-t^2 + 4t)\mathbf{i} + (\frac{1}{6}t^3 - \frac{1}{2}t^2 + t)\mathbf{j}$
(c) 2 s
9 (a) 46.6 N at 014.9°
(b) $(\frac{2}{3}t^{\frac{3}{2}} + t)\mathbf{i} + (t^2 - 3t + 2)\mathbf{j}$
10 (a) $(4t - t^2 - 3)\mathbf{i} + (2t^3 - 2)\mathbf{j}$
(b) Yes, at $t = 1$ both components of the velocity are zero.
(c) $(2t^2 - \frac{1}{3}t^3 - 3t)\mathbf{i} + (\frac{1}{2}t^4 - 2t)\mathbf{j}$
(d) $t = \frac{1}{2}$
11 9 m
12 (a) $\mathbf{a} = 3\mathbf{i} - 5\mathbf{j}$
(b) $\sqrt{125} = 5\sqrt{5} = 11.2$ m
13 (a) $2t^2 - 8t + 6$
(b) $2\frac{2}{3}$ m
14 $\mathbf{a} = 2\mathbf{i} - 4\mathbf{j}$
$2\sqrt{5}\ \text{ms}^{-2}$

CHAPTER 3

EXERCISE 3A (Page 57)

1 (a) 0.2 m
(b) −0.72 m
(c) 0.275 m
(d) 1.19 m
(e) 0 m
(f) −0.92 m
(g) 0.36 m
(h) 0.47 m
2 2.18 m from 20 kg child
3 4.2 cm
4 4680 km
5 0.92 m
6 3.33 mm from centre
7 2.95 cm
8 1.99 kg
9 42 kg
10 $\dfrac{m_2 l}{(m_1 + m_2)}$ from m_1 end
11 (a) 3.35 m, tips over
(b) 4.55 tonnes
(c) $L(l - d) < Md + C(a + d)$, $C(a - d) < Md$
(d) $\dfrac{2Mad}{(l - d)(a - d)}$

EXERCISE 3B (Page 65)

1 (a) (2.3, −0.3)
(b) (0, 1.75)
(c) $(\frac{1}{24}, \frac{1}{6})$
(d) (− 2.7, − 1.5)
2 $(5, 6\frac{1}{3})$
3 (a) (20, 60)
(b) (30, 65)
(c) (30, 60)

4 23 cm
5 (a) (1.5, −1.5)
(b) (1.5, −2.05)
(c) (1.68, −2.5)
6 (a) (28, 60)
(b) (52, 60)
(d) 40 cm
7 (a) (i) (10, 2.5)
(ii) (12.5, 5)
(iii) (15, 7.5)
(iv) (17.5, 10)
(v) (20, 12.5)
(b) 5
(c) $(9 + n, 2.5n)$, 11
(d) 102.5 cm
8 (a) 0.2 cm below O
(b) 9.1°
9 (a) $(0.5a, 1.2a)$
(b) 3.9°
(c) $2m$
10 (a) 2.25
(b) 0.56 m; the hole is larger than the metal plate.
(c) 0.40, $(\frac{1}{2}, 1\frac{1}{2})$

EXERCISE 3C (Page 69)

1 (a) 7.62 cm
(b) 27.7°
2 (a) 8.76 cm
(b) 23°
3 (a) On the line of symmetry, 8.91 cm from AB.
(b) 31°
4 (a) $\frac{19a}{28}$
(b) $\frac{13b}{28}$
5 (a) $\frac{1 + 6k}{1 + 4k}a, \frac{1 + 2k}{2(1 + 4k)}a$
(b) 3
6 (a) 3.59 cm
(b) 4.54 cm
(c) 23.9°
(d) 0.804
7 (a) 2.48 cm
(b) 4.52 cm
(c) 20.8°
8 (a) $\frac{1}{2}a$ from AB, $\frac{23b}{42}$ from AD.
(b) $\tan^{-1}\left(\frac{2b}{21a}\right)$
(c) $2m$

9 (a) 0.417 cm
(b) 5.23 cm
(c) 12.1°
(d) The normal reaction will be at 12.1° to the vertical, but the total reaction must be vertical.
10 (a) 5.2 cm
(b) 4.2 cm
(c) All the masses are on or to the right of AB, so the centre of mass cannot be directly below B.
(d) 0.3
11 (b) −1.1
12 (a) $\frac{19a}{15}$
(b) $\frac{7M}{45}$
13 (a) 6.86 cm
(b) 32.1°

CHAPTER 4

EXERCISE 4A (Page 82)

1 (a) 2500 J
(b) 40 000 J
(c) 5.6×10^9 J
(d) 3.7×10^{28} J
(e) 10^{-25} J
2 (a) 1000 J
(b) 1070 J
(c) 930 J
(d) None
3 (a) 4320 J
(b) 4320 J
(c) 144 N
4 (a) 540 000 J, no
(b) 3600 N
5 (a) 500 000 J
(b) 6667 N
6 (a) (i) 5250 J
(ii) −13 750 J
(b) (i) 495 250 J
(ii) 476 250 J
7 (a) 64 J
(b) Dissipated
(c) 64 J
(d) 400 N
(e) 89.4 ms^{-1}
8 (a) 3.146×10^5 J
(b) 8.28×10^3 N

(c) Dissipated as heat and sound
(d) Some of work is dissipated.

9 $18.6\,\text{ms}^{-1}$

10 (a) 240 N
(b) 5.5 m
(c) 1320 J
(d) 270 J
(e) 960 J, 90 J

EXERCISE 4B (Page 91)

1 (a) 9.8 J
(b) 94.5 J
(c) −58.8 J
(d) −58.9 J

2 (a) −27.44 J
(b) 54.9 J
(c) −11.8 J

3 17.6 J

4 23 300 J

5 (a) 154 000 J
(b) 20 000 J
(c) Distance moved against gravity is $200\sin 5^\circ$
(d) 138 000 J

6 (a) (i) 1500 J
(ii) 280 J
(b) (i) $15.6\,\text{ms}^{-1}$
(ii) $16.1\,\text{ms}^{-1}$

7 (a) 2120 J
(b) (ii) The same

8 (a) (i) 1700 J
(ii) $8.7\,\text{ms}^{-1}$
(b) (i) Unaltered
(ii) Decreased

9 (a) 1750 J
(b) 1750 J, $8.37\,\text{ms}^{-1}$
(c) 50°
(d) It is always perpendicular to the motion.

10 (a) 154 J
(b) $153.7 - 1.96x$
(c) $0 \leqslant t \leqslant 4$
(d) $27.7\,\text{ms}^{-1}$
(e) 58.8 m

11 (a) 34 300 J, 21 875 J
(b) 248.5 N
(c) 5061 N

12 (a) $9.8\,\text{ms}^{-2}$
(b) $1.47(10t - 4.9t^2), 0 \leqslant t \leqslant 2.04$
(c) 5.1 m, 10**i**
(d) 10**i** + 10**j**, $14.1\,\text{ms}^{-1}$, 15 J
(e) No air resistance; no

13 (a) $109\,\text{ms}^{-1}$
(b) 115 N
(c) The heavier (relatively less affected by resistance).

14 (a) 12 J, 8.76 J
(b) 3.24 J
(c) 0.217 N
(d) $14.1\,\text{ms}^{-1}$

15 (a) 592 J
(b) 759 000 J
(c) 211 W

16 (a) 5 m
(b) $6.26\,\text{ms}^{-1}$
(c) 8 m
(d) Yvette is a particle and remains upright. No air resistance.

EXERCISE 4C (Page 98)

1 (a) 308.7 J
(b) 37 044 J
(c) 10.3 W

2 (a) 2352 J
(b) 1176 W
(c) 1882 W, 0 W, 2822 W

3 (a) 31 752 J
(b) 16 200 J
(c) 1598 J
(d) 1332 N
(e) Power = 1598 J

4 (a) 703 N
(b) Mass of car

5 576 N

6 245 kW

7 (a) 560 W
(b) 168 000 J

8 (a) 1250 J
(b) 209 W

9 (a) $20\,\text{ms}^{-1}$
(b) $0.0125\,\text{ms}^{-1}$
(c) $25\,\text{ms}^{-1}$

10 (a) 1.6×10^7 W
(b) $0.0025\,\text{ms}^{-2}$
(c) $5.7\,\text{ms}^{-1}$

11 (a) 16
(b) 320 N
(c) 6400 W
(d) (i) $1.98\,\text{ms}^{-2}$
(ii) $1.12\,\text{ms}^{-2}$

EXERCISE 4D (Page 100)

1 0.6 N
2 1.34 m
3 11 300 kW
4 12.7 ms^{-1}
5 (a) 314 W
(b) No resistance to motion; no losses to friction accelerating the sand.
6 (a) 2.48 ms^{-1}
(b) 0.31 m
7 (a) 2 ms^{-1}
(b) 3.16 m
8 (a) 35.2 kJ decrease.
(b) Losses to air resistance.
9 (a) 294 W
(b) $P = 294 + \frac{5}{6}R$
(c) $P = 588,\ R = 353$
10 (b) 0.5 ms^{-2}
(c) 27.6 ms^{-1}
11 86.5 kW
12 (a) 30 N
(b) 208 W
(c) 16.9 N
(d) 168 W
13 (a) 1.82 ms^{-2}
(b) 1800 N
(c) 28.1 m

CHAPTER 5

EXERCISE 5A (Page 106)

1 (a)
(i) (ii) (iii)

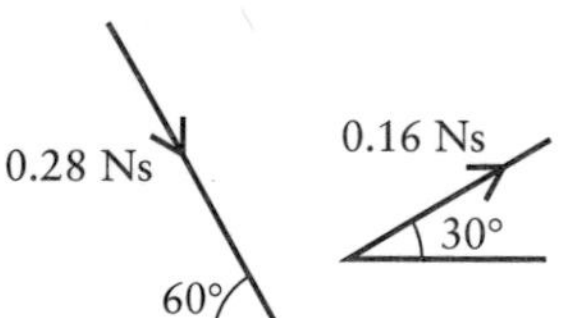

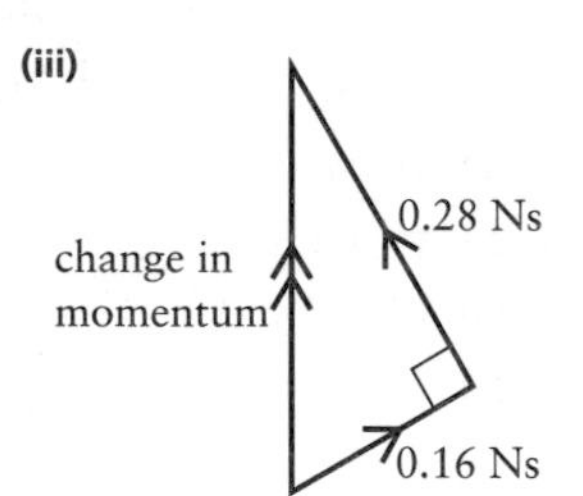

(c) 1.75i − 3.03j, 1.73i + j
(d) 0.32 Ns at 90.3°. There is hardly any change in momentum parallel to the cushion.
2 (a) −20i
(b) 5.1i + 1.05j
(c) Fatima
3 (a) (i) 0.048 Ns right
(ii) 0.048 Ns left
(b) 0.432 N

EXERCISE 5B (Page 111)

1 (a) $\frac{2}{3}$
(b) 1.44 ms^{-1}
(c) $\frac{3}{4}$
(d) 3.2 ms^{-1}
2 (a) 0.3
(b) 0
(c) 0.95
(d) 1
3 (a) 0.8
(b) 1.62 Ns
(c) 2.43 J
4 (a) 4.43 ms^{-1}
(b) 3.98 ms^{-1}
(c) 0.9
(d) 0.149 J
(e) 0.149 J
(f) 0.656 m
5 (a) 1 ms^{-1}
(b) 0.2
(c) 450 Ns
(d) 900 J
6 (a) 20 ms^{-1}
(b) $\frac{3}{4}$
(c) 1750 Ns
(d) 4375 J
(e) low
7 (a) (ii) 2.5 ms^{-1}, 3.5 ms^{-1}
(iii) 3.75 J
(b) (ii) −0.5, 2.5 ms^{-1}
(iii) 33.75 J
(c) (ii) 1.2 ms^{-1}, 1.2 ms^{-1}
(iii) 4.8 J
(d) (ii) −1 ms^{-1}, 2 ms^{-1}
(iii) 0 J
(e) (ii) −0.5 ms^{-1}, 1 ms^{-1}
(iii) 2.25 J
(f) (ii) 2 ms^{-1}, 4 ms^{-1}
(iii) 96 J
8 (a)

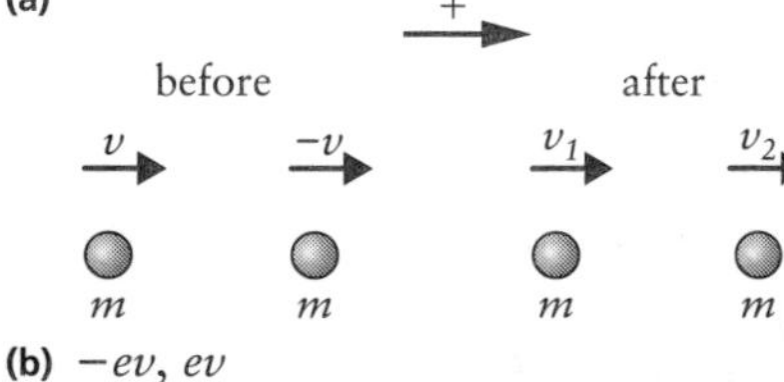

(b) $-ev,\ ev$
9 (b) 2 ms^{-1}
(c) $\frac{2}{3}$
(e) $m_1 \leqslant m_2$
(f) $e : 1,\ 1$

10 (a) $\sqrt{\frac{2h}{g}}, \sqrt{2gh}$

(b) e^2h

(c) he^{2n}

(d) $\sqrt{\left(\frac{2h}{g}\right)}(1 + 2e)$,

$\left(\sqrt{\frac{2h}{g}}\right)(1 + 2e + \ldots + 2e^{n-1})$

(f) $\dfrac{h(1 + e^2)}{(1 - e^2)}$

Exercise 5C (Page 118)

1 $\frac{13}{32}, \frac{15}{32}, \frac{9}{8}$

2 $-\frac{47}{16}, \frac{7}{16}$

3 (a) $\dfrac{11u}{36}, \dfrac{5u}{9}$

(b) $-\dfrac{5u}{117}, \dfrac{175u}{468}$

(c) Yes, as B is moving towards A.

4 (a) $0.15u$, $0.85u$

(b) $1.275mu$

(c) $-0.339u$, $0.064u$

(d) $0.441mu^2$

5 (a) 0.8

(b) $-\dfrac{u}{5}, \dfrac{12u}{25}$

Exercise 5D (Page 119)

1 (a) $-\mathbf{i} + 6\mathbf{j}\,\text{ms}^{-1}$

(b) 36.0°

2 (a) $-4.5\mathbf{i} + 0.6\mathbf{j}$ Ns

(b) 130.0°

(c) 18.9 J loss

3 $0.35\mathbf{i}$ N

4 (c) $\dfrac{48mu^2}{7}$

5 (b) $\frac{11}{14}$

(c) $3.67mu^2$

6 0.323

7 (a) $\frac{u}{3}$

8 (b) $-\frac{6}{175}(1 + e)u$

9 (a) $-\frac{u}{9}, \frac{5u}{9}$

(c) $0.447mu^2$

10 $-0.17u$, $-0.02u$

11 (a) 5.83 Ns

(b) 31°

(c) 35 J

12 (a) $\frac{1}{3}(1 - 2e)u$, $\frac{1}{3}(1 + e)u$

(b) $0 \leqslant e < \frac{1}{2}$

(c) Speed of B is $\frac{1}{9}(1 + e)(1 - 2e)u$

13 (a) 14.8 Ns

(b) $22.5\,\text{ms}^{-1}$

(c) 48.1°

(d) Air resistance, spin

Chapter 6

Exercise 6A (Page 129)

1 (a) 6 Nm

(b) −10.7 Nm

(c) 23 Nm

(d) 0

(e) −4.24 Nm

(f) 4.24 Nm

2 David and Hannah (by radius × 0.027 Nm)

3 (a) 5915 kg

(b) $4532\sec\theta$ kg

4 (a) 42.4 N

(b) 27.7 N

(c) 30.1 N

5 (a) $T\cos30°$, $T\sin30°$

(c) 30 Nm

(e) 8.04 N, 15.36 N

(f) (i) 33.7°

(ii) 3.23 m

6 (a) 1405 N

(b) 638 N, 1612 N

(d) (i) Jib stays put, $T = 0$.

(ii) A falls.

7 (a)

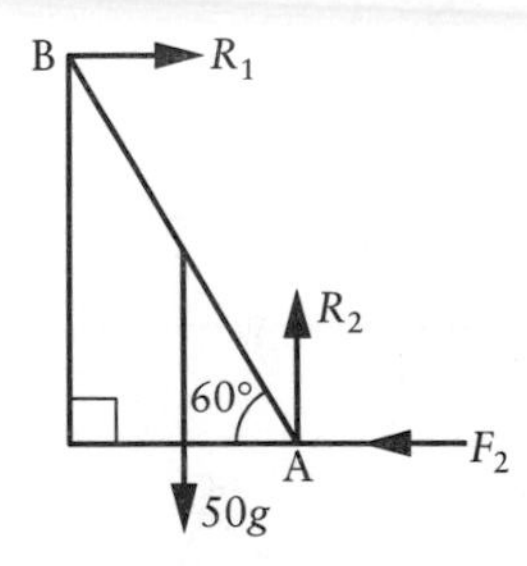

(b) 0

(c) 141 N

(d) 141 N, $\mu \geqslant 0.289$

8 (a)

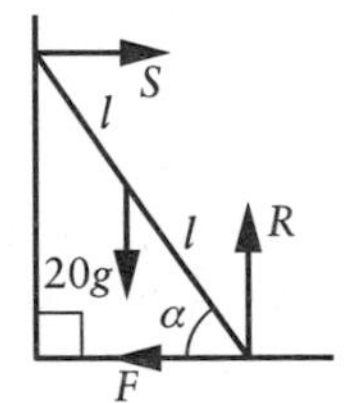

(i) (b) 56.6 N, 56.6 N, 196 N

(c) 0.29

(ii) (b) 98 N, 98 N, 196 N

(c) 0.5

9 (a)

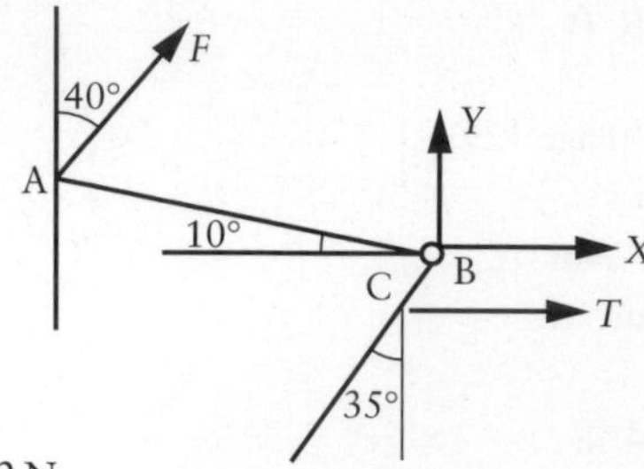

(b) 162 N

(c) 61.7 N

10 (b) 1 600 000 Nm

(c) 6830 N, 3830 N

11 (a)

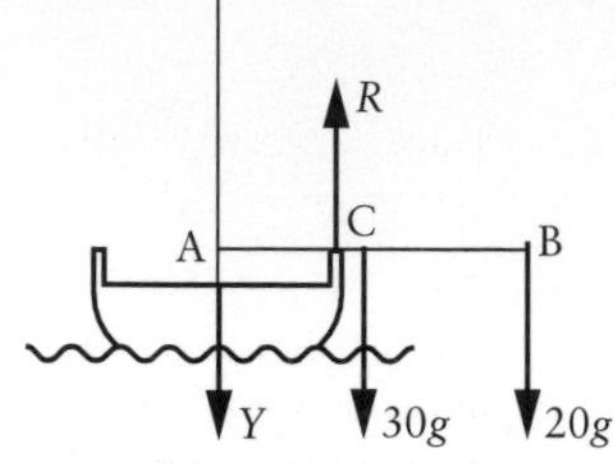

(b) $80g = 784$ N

(c) $30g = 294$ N vertically down

(d) $17.5g$ N

12 (a)

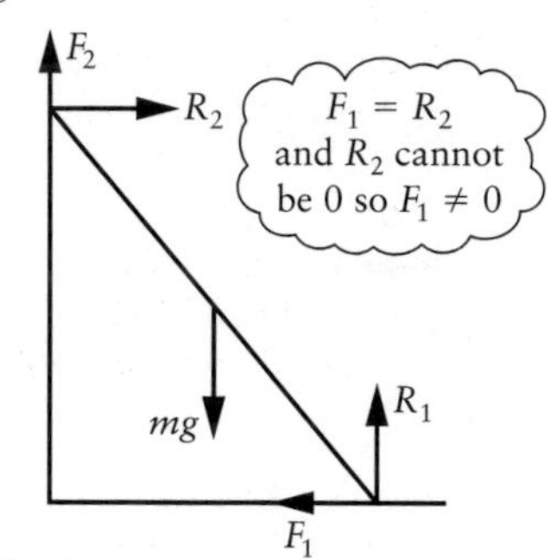

(b) (ii) $32g$ N

13 (a)

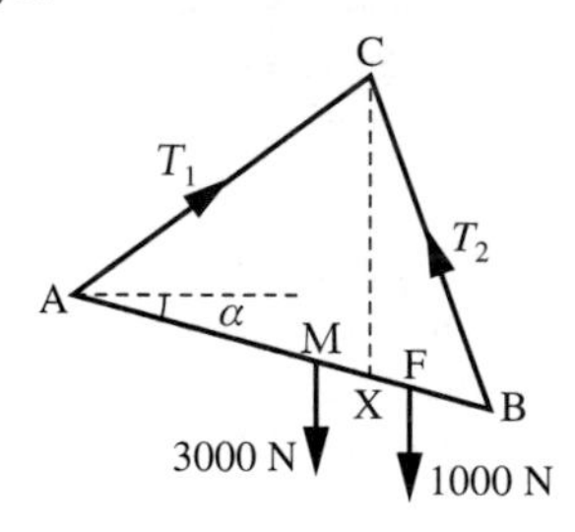

(b) 0.25 m, 0.75 m; 4 m, 3.6°

(c) 2290 N, 2700 N

(d) 10.6°

14 (a) 19.4°

(b) $\tan\phi = 2\tan\theta$

Exercise 6B (Page 134)

1 164 N

2 70.2°

3 (a) 184 N

(b) 67.0 N

4 (a) 70 N

(b) 321 N

5 (a) 462 N

(b) 7.35 N down

6 (a) 13.5°

(b) 14.8 N, 36.2 N

(c) 0.409

7 (a) 61.25 N

(b) 165 N

(c) 53 N

(d) $\mu \geqslant 0.321$

8 (a) 59.0°

(b) $0.402M$

9 (b) 57.4°

10 (a) 0.182

(b) 0.662 kg

11 $\frac{1}{4}$

12 (b) 877 N

(c) By making the tension at D equal to the weight of the load.

13 (b) 68.2°

(c) 0.646

INDEX